"INTENSELY GR~~IP~~ P9-CDI-898
—*Winnipeg Free Press* (a Best Book of the Year)

"PROFOUNDLY IMPORTANT."
—*Kirkus Reviews* (starred review; a Best Book of the Year)

"POWERFUL."
—*Christian Science Monitor*
(a 10 Best Nonfiction Books of the Year selection)

"FASCINATING."
—*New York Times Book Review* (Editors' Choice)

"RIVETING."
—*San Francisco Chronicle* (a Best Book of the Year)

"IRRESISTIBLE. START, AND THERE'S NO LETTING GO."
—*Star Tribune* (Minneapolis, MN)

"A MASTERPIECE."
—Robert Kurson

"A TOUR DE FORCE."
—Douglas Preston

"NARRATIVE NONFICTION AT ITS FINEST."
—Jon Huntsman

"A BOOK THAT CAN SAVE LIVES."
—Nicholas Carr

A BOOKS FOR A BETTER LIFE AWARDS FINALIST

NEW YORK TIMES BESTSELLER

Praise for *A Deadly Wandering*

"Compassionate and persuasive. . . . As instructive social parable . . . it deserves a spot next to *Fast Food Nation* and *To Kill a Mockingbird* in America's high school curriculums. To say it may save lives is self-evident." —*New York Times Book Review* (Editor's Choice)

"Matt Richtel's keen and elegantly raw—like a tooth-cracklingly crisp photograph that bleeds at the edges—story surrounding this disaster is not just a morality tale about texting and driving, but also a probe sent into the world of technology, examining the way it is outstripping our capability to keep up with it, and how we as a culture are feeding bullets into the techno-gun and playing with it."

—*Christian Science Monitor* (a 10 Best Nonfiction Books of the Year selection)

"A masterful book. . . . Irresistible. Start, and there's no letting go. . . . A well-paced emotional story of destruction, atonement, and redemption; a hero's tale. Richtel's masterful telling makes him a literary hero of sorts, too." —*Star Tribune* (Minneapolis, MN)

"Matt Richtel's riveting book is narrative nonfiction at its finest. A well-written true-life account of tragedy, redemption, and the challenges of keeping pace with the march of technology. This book should be placed in every school and legislative chamber in the country."

—Jon Huntsman, former governor of Utah

"An arresting tale." —*San Francisco Chronicle* (a Best Book of the Year)

"This is a gripping book about a young man, his torment, and his redemption, after a distracted-driving disaster. It will make you an attentive driver—*or else*. *A Deadly Wandering* is human drama and the latest knowledge about obsessive technology woven together in memorable style."

—Ralph Nader, author of *Unsafe at Any Speed*

"An intensely gripping, compelling, and sobering retelling of the accident and its painful emotional and legal aftermath. Reggie emerges as a complex, sympathetic and tragic figure, one with whom almost any driver can, with no small unease, identify. . . . *A Deadly Wandering* gives the potentially lethal risks of the digital age a very human face—one which we can, if we're honest, readily see in the mirror."

—*Winnipeg Free Press* (a Best Book of the Year)

"*A Deadly Wandering* is the perfect companion to Matt Richtel's Pulitzer Prize–winning reporting on how technology impacts our attention, decisions, and daily lives. What's more, this book does that most amazing of feats: it makes cutting-edge scientific research feel relevant to the choices we make every time we get in a car, sit at a desk, or talk to our friends and family."

—Charles Duhigg, author of *The Power of Habit*

"*A Deadly Wandering* delivers an important message through a gripping and moving narrative of tragedy and redemption. Richtel has complete mastery of this complex topic and palpable compassion for the people who shared their experiences with him. . . . A page-turner with a lesson we should all heed."

—*Idaho Statesman*

"Through an elemental story of death and atonement, Matt Richtel explores the sometimes lethal tension that exists between our multitasking gadgets and our uni-tasking minds. *A Deadly Wandering* is more than a page-turner. It's a book that can save lives."

—Nicholas Carr, author of *The Shallows*

"A compelling, highly emotional, and profoundly important story."

—*Kirkus Reviews* (starred; a Best Book of the Year)

"A masterpiece of reporting, insight, and empathy. *A Deadly Wandering* is a journey into our onrushing everyday world, where technology is colliding with the limits of the human brain. In writing it, Matt Richtel has given us a beautiful, cautionary tale that reads like a novel, and that we disregard at our risk."

—Robert Kurson, author of *Shadow Divers*

"Illuminates the perils of information overload. . . . Raises fascinating and troubling issues about the cognitive impact of our technology."

—*Publishers Weekly*

"Americans are addicted to their technology, putting us on a modern-day collision course with very real consequences. Matt Richtel brilliantly tells the story of the aftermath of a deadly distracted driving crash. His portrait is riveting. I could not stop reading, and neither will you."

—Ray LaHood, former U.S. secretary of transportation

"*A Deadly Wandering* is part human tragedy, part fascinating neurological exploration."

—CBC (Canada)

"One of the most important books published in 2014. . . . A rising star in the crime fiction genre—with such terrific tech exposé novels as *The Cloud* and *The Devil's Plaything*—Richtel has combined his acute research skills with his storytelling mastery to produce a nonfiction book with the page-turning power of a thriller."

—*Connecticut Post*

"Gripping and compelling, *A Deadly Wandering* serves up a kaleidoscopic view of a searing tragedy, one that sprang from our fixation with communication gadgets. Matt Richtel writes like a powerful novelist, deftly crossing many levels from the secret lives of teens to police investigations to the psychology of attention. His probe into a moment of fatal distraction yields a poignant, deeply moving story. Readers will be rewarded with acute insight into the human mind and heart. The book's intertwined story will surely touch all who read it, and may even save your life."

—James E. Katz, PhD, executive director,
Center for Mobile Communication Studies, Boston University

"Richtel is an unflaggingly energetic writer. . . . [*A Deadly Wandering* is] gripping reading [that raises] many powerful and complex questions."

—*Open Letters Monthly: Arts and Literature Review*

"Riveting. . . . A must-read."

—*Steamboat Today* (Colorado)

ALSO BY MATT RICHTEL

The Doomsday Equation

The Cloud

Floodgate

Devil's Plaything

Hooked

A DEADLY WANDERING

A Mystery,
a Landmark Investigation,
and the Astonishing Science
of Attention in the Digital Age

MATT RICHTEL

WM
WILLIAM MORROW
An Imprint of HarperCollins*Publishers*

HarperCollins books may be purchased for educational, business, or sales promotional use. For information please e-mail the Special Markets Department at SPsales@harpercollins.com.

A hardcover edition of this book was published in 2014 by William Morrow, an imprint of HarperCollins Publishers.

FIRST WILLIAM MORROW PAPERBACK EDITION PUBLISHED 2015.

Image on title page and part title pages copyright by April70/Shutterstock, Inc.

Designed by Lisa Stokes

Library of Congress Cataloging-in-Publication Data has been applied for.

ISBN 978-0-06-228407-5

17 18 19   ov/rrd   20 19 18 17 16 15 14 13 12 11

For my family

We have Paleolithic emotions, medieval institutions, and godlike technologies.

E. O. WILSON

CONTENTS

A DEADLY WANDERING

PROLOGUE

"ARE YOU COMFORTABLE, REGGIE?"

"Yep."

Reggie Shaw lies on a medical bed, his head inches from entering the mouth of a smooth white tube, an MRI machine. He's comfortable, but nervous. He doesn't love the idea of people peering into his brain.

Next to the machine stands a radiology technician in blue scrubs, her hair pulled tightly into a bun. She scans the room to make sure there are no errant pieces of metal. The MRI, with sixty thousand times the strength of the earth's magnetic force, is a kind of irresistible magnet. A small pair of scissors, if accidentally left out, could be sucked across the room into the tube at forty miles an hour.

Reggie, twenty-six, has removed his clothes and left outside his keys, and the iPhone he keeps so regularly in his left front pocket it leaves a faint outline on the jeans. With his head at the edge of the machine, he wonders whether the permanent retainer on his bottom teeth, the product of a particularly nasty clash in a recreational football game in high school, could get yanked through his head. The technician, Melody Johnson, assures Reggie he'll be okay.

She walks to the left of the machine and from a table lifts an odd-looking helmet, a cross between something that might be worn by an astronaut and Hannibal Lecter.

"I'm going to place this over your head." She fits the white helmet over Reggie's face, clipping the sides down to the bed. Inside the helmet, there's a small mirror. Images can be projected into it in such a way that Reggie, lying flat on his back, stuffed in the tube, will be able to see them.

The hum of the whirring machinery is so loud that Reggie wears earplugs. The MRI works by sending massive amounts of magnetic energy into the person's body. This excites hydrogen atoms, which are in heavy concentrations in water and fat. As the atoms begin to settle back down from their briefly excited state, they give off a radio frequency, not unlike that of an FM station. Then the computer picks up the signal and translates it into physical images—a map, or topography, of the inside of the body. The technology isn't great for looking at hard structures, like bone, but it's extraordinary at imaging soft tissue, like organs. It's an unprecedented tool for looking at the brain.

When Reggie was little, he dreamed he'd play college basketball, or maybe coach. He'd have a family, for sure, but not just for its own sake; jock though he might have been, he was a romantic who wanted to fall in love, and to be in love. He hoped most of all to go on a Mormon mission. Then, one rainy morning in September 2006, while Reggie was driving to work on a mountain pass, life took a tragic, deadly turn. There was an accident, or so it seemed. Maybe it was just a moment of inattention, or something more insidious. Exactly what happened that last day of summer was not yet clear.

Two men were dead, leaving behind extraordinary grief—and a mystery. The case attracted a handful of dogged investigators, including a headstrong Utah State Trooper. He became convinced that Reggie had caused the wreck because he'd been distracted by his cell phone, maybe texting. He pursued a stubborn probe, a lonely one at first, looking for evidence and proof of Reggie's wrongdoing, but discovering only one obstacle after another. And, later, there was a victim's advocate, a woman named Terryl Warner. She had survived a terrible childhood, one that toughened her and forged an uncompromising sense of duty she used to pursue justice for the crash's victims.

For his part, Reggie claimed not to remember what caused the crash. Then, as the evidence emerged, Reggie denied it, deceived himself, and was reinforced in his denial and deception by his most loving friends and family. Some members of the community, while sympathetic to the victims, couldn't understand the fuss. So what if he'd looked at his phone, or texted—haven't we all been distracted behind the wheel? Who knew that was so wrong? The law was no help: Nobody in Utah had ever been charged with such a crime.

The accident became a catalyst. It spun together perspectives, philosophies, and lives—those of Reggie and his advocates, and Terryl and the other pursuers, including, ultimately, prosecutors, legislators, and top scientists. It forced people to confront their own truths, decades-old events, and secrets that helped mold them and their reactions—in some cases conflicted and in others overpowering—to this modern tragedy.

And this maelstrom of forces left behind a stark reality. The tragedy was the product of a powerful dynamic, one that elite scientists have been scrambling to understand, even as it is intensifying. It is a clash between technology and the human brain.

Broadly, technology is an outgrowth of the human mind. It is an extraordinary expression of innovation and potential. Modern-day machines serve us as virtual slaves and productivity tools. The value of such technology is inarguable in every facet of life—from national security and medicine to the most basic and intimate, like the way far-flung family and friends are nurtured and connected through miniature, ubiquitous phones; email that travels thousands of miles in seconds; or Skype and FaceTime. Fundamentally, the extraordinary pace at which consumers adopt these programs and gadgets is not the product of marketing gimmicks or their cool factor but because of their extraordinary utility. They serve deep social cravings and needs.

At the same time, such technology—from the television to the computer and phone—can put pressure on the brain by presenting it with more information, and of a type of information, that makes it hard for us to keep up. That is particularly true of interactive electronics, delivering highly relevant, stimulating social content, and with increasing speed. The onslaught taxes our ability to attend, to pay attention, arguably among the most important, powerful, and uniquely human of our gifts.

As Reggie's story unfolded, it illuminated and contributed to a thread of science dating to the 1850s, when scientists began to measure the capacities of the human brain—how we process information, how quickly, and how much of it. Prior to that time, the conventional wisdom was that people could react instantly. The idea was that the human brain was "infinite." Machines began to change that thinking. Compared to, say, guns or trains or the telegraph, people's reaction times didn't seem so instant. Technology was making us look slow. But it was also allowing scientists to study the brain, creating an interesting trade-off; machines highlighted the limitations of the brain, threatened to stress our processing power and reaction time to the breaking point, but they also allowed scientists to understand and measure this dynamic.

Then, around World War II, modern attention science was born, also prompted by people's relationships to technology. A generation of pioneering researchers tried to figure out how much technology pilots could handle in the cockpit, and tried to measure when they became overloaded, and why. Or why radar operators, looking at cutting-edge computer displays, were sometimes unable to keep up with the blips that showed Nazi planes.

In the second half of the twentieth century, high tech moved from the military and government to the consumer. First came radio, and then television (the demand for it growing explosively from 3.6 million sold in the United States in 1949 to an average of three per American home in 2010). Computers followed; the first mouse pioneered in the early 1960s, the personal computer a decade later. By the 1980s, the commercial mobile phone exceeded by orders of magnitude the capability of the world's greatest military computer in World War II. And within a few years, it would be right there in the pocket.

The developments were swift, the acceleration described by Moore's law, which, in essence, talks of computer processing power doubling every two years. There was something else, a principle less celebrated than Moore's law but of equal significance when it comes to understanding what is happening to the human brain. The axiom is called Metcalfe's law. It was codified in the early 1990s, and it defines the power of a computer network by the number of people using it.

More people, more communication, more value.

More pressure.

As networks became more populated and powerful, they added a huge wrinkle in the demand for attention by turning computers into personal communication devices. The technology was delivering not just data but information from friends and relatives—communications that could signal a business opportunity or a threat, an overture from a mate or a potential one. As such, the devices tapped into deep human needs—with increasing speed and interactivity. It was not just pure social communications, but video games, news, even shopping and consumption, a powerful, personalized electrical current connecting all of us, all of the time. This was the marriage of Moore and Metcalfe—the coming together of processing power and personal communications—our gadgets becoming faster and more intimate. They weren't just demanding attention but had become so compelling as to be addictive.

The modern attention researchers, walking a path laid down by their forebears 150 years earlier, asked a new question: Was technology no longer the slave, but the master? Was it overtaking our powers of attention? How could we take them back? It wasn't just a question of life-or-death stuff, like the stakes for pilots in World War II. Now there were subtler tensions, the concept that nips and cuts at attention in the cubicle can take a persistent and low-grade toll on productivity, or in schools on focus, or at home on communication between lovers and parents and children. Would it hinder memory and learning rather than enhance it?

Past technological advances, from the printing press to the radio and television, had invited questions about their unintended consequences and possible negative side effects. But many scholars agreed that these latest breakthroughs, taking full form only in the last decade, marked a difference in our lives in orders of magnitude.

Technology was exploding in complexity and capability. How could we keep up?

Reggie Shaw could not—keep up. He could not conceive of the larger dynamic, even the crisis, that had enveloped him. So maybe it's no wonder he couldn't grasp what had happened; perhaps this confusion prompted him to deceive himself and lie to others. Or was he less innocent than he was letting on? In any case, after being pressed by science and common sense, he no longer could keep the truth at bay and he recognized what he'd done, and he changed, completely. He became the unlikeliest of evangelists, a symbol of reckoning. And he began to transform the world with him. Broadly, his story, and that of others around him, became an era-defining lesson in how people can awaken from tragedy, confront reality, address even smaller daily dissonance, and use their experiences to make life better for themselves and the people around them. And their journey showed how we might come to terms with the mixed blessing of technology. For all the gifts of computer technology, if its power goes underappreciated, it can hijack the brain.

Along the way, Reggie's defenders and antagonists alike came to see themselves in the young man, a projection of how they would've handled themselves, or should. His attention, ours, is so fragile. What happened to him could happen to anyone, couldn't it? Does that make him, or us, evil, ignorant, naive, or just human?

Is his brain any different from ours?

Ms. Johnson, the technician, hands Reggie two little plastic devices, gray, looking like primitive video game joysticks. She tells him that the gadgets have buttons he'll be asked to press when certain images appear in the mirror. They're going to see what Reggie's brain looks like when he tries to pay attention.

"I'm going to put you in slowly, Reggie," says Ms. Johnson. "Is that okay?"

Reggie clears his throat, a sign of his assent, an exhalation of nerves. He disappears into the tube.

PART ONE

COLLISION

CHAPTER 1

REGGIE

IN EARLY JUNE 2006, nineteen-year-old Reggie Shaw sat in the back-seat of a Chevy Tahoe heading north under a big, cloudless Utah sky. His father, Ed, a machine-shop manager, was crying quietly as he drove the white sport-utility vehicle. In the passenger seat, Reggie's mother, Mary Jane, sobbed.

Reggie was her little boy, her baby, at least until his little sister came along when Mary Jane was forty. Among her brood of six, Reggie was the quiet charmer, a peacemaker, sensitive with a dry wit, both athletic and awkward, honest. This time to a fault.

A day earlier, Reggie had been sitting in a classroom in Provo at the Missionary Training Center. He was surrounded by eager teen Mormons, each preparing to embark on a mission, Reggie's lifelong dream. He'd recently returned from his freshman year at a small Mormon college in Virginia, where he'd played basketball, and he was committed to taking the Mormon message to Winnipeg. But a secret nagged at him. He went to his president at the training center, and he confessed: He'd recently had sex with his girlfriend, Cammi.

The fact that he'd previously lied about their coupling, and hadn't done the spiritual work to put it behind him, ruled out his participation in the mission. Most horrifying to Reggie was the knowledge that the Church would soon phone his parents. The family lived in Tremonton, in the northernmost part of Utah. It had some of the heaviest concentration of Mormons in the state and, by extension, in the world. When someone came home early from a mission, everyone in the community knew about it, and people would suspect the reason. Even though the Shaws were well regarded, with deep roots, Reggie felt he'd marked not just himself, but his family.

"It was difficult for them to drop me off knowing I wouldn't be back for two years," he says, looking back. "It was much more difficult to pick me up." His dad was a quiet man, particularly if you didn't know him well, someone who ached for his children when they hurt, even if he couldn't quite express it. This was the first time Reggie could remember seeing him cry.

Reggie uttered hardly a word as they wound their way through Salt Lake City on I-15 North. It was nearly seven p.m. The sun, falling in the west, to the left of the Chevy, snuck into the car at an insidious angle, causing Reggie to squint. He had a short haircut, leaving a touch of bangs in the front. His young face usually projected kindness, approachability, but now held heavy weight he had no language to express.

In the distance, to the right of the car, in the east, rose the Wasatch Range of the Rocky Mountains. The imposing jagged peaks put the topography out of balance, almost tilting the land in their direction; the mountains had a gravity that helped define the state, outline it, just as they would come to define Reggie.

The family continued north, passing through Salt Lake City, things and places blurring—the auto mall, McDonald's, Best Buy, the exit for the University of Utah. As they drove, Reggie thought about Cammi, and wondered what would become of them, and her, of him.

Less than an hour later, they arrived at the two-story, red-brick home in Tremonton, the town where Reggie had grown up, and his mom had, too. Her family had raised sugar beets, cattle, hay, and corn. Lots of land, few people. Everyone still knew everyone. By 2006, there were fewer than six thousand residents. Down the block from the Shaws lived the town's mayor; kitty-corner from the mayor lived their LDS bishop. They were all within walking distance from Reggie's elementary, middle, and high schools, from the Little League field his dad had helped care for, and the recreation center where Reggie first learned to play his great passion.

A few days after they returned home from the failed mission, on a Sunday, Reggie went to church. Mary Jane felt a cascade of emotions. Heartbroken, embarrassed. "If you come home like that, it's almost like a disgrace," Reggie's mom says now. "But he walked right back into church. He never faltered.

"I was very proud of him for having the courage to do that."

That summer, Reggie took a painting job at Wall to Wall, a company in Logan, the veritable big city to the east, over and then around the first batch of mountains, past Chocolate Peak and Scout Peak. Every morning, it was the same thing; Reggie was up and out of the house by six a.m., he'd drive the Tahoe north through town in the dark, then take a right at Valley View Drive, where things got wide open, and then accelerate up the crest into the foothills.

He lived in the room he'd once shared with his older brother Nick. It was all boy; Chicago Cubs wallpaper covered the bottom third of the wall. There was a poster of Reggie's favorite basketball player, Reggie Miller, but the star's bio was covered up by a picture of Jesus, looking serene, wearing a white shirt with a red robe over it.

Reggie tried to reconnect with Cammi. "She was the one, man. That's what I thought." She didn't share his resolve. He couldn't quite figure out why things weren't working, her periodic distance. Then one day, she stopped taking his calls. He couldn't get ahold of her. Then she reappeared. "She called me up and said she'd gotten engaged to someone else."

By September, he'd developed a rhythm. Painting, trying to figure out what would come next, playing recreational hoops and video games, and forming a new friendship, with Briana Bishop. Still just a friendship, but with potential.

Most of all, he was doing the spiritual work to cleanse his transgression. He was determined to get square with his Church and Maker so he could embark again on a mission, sometime the following year. It was not the path that Reggie had once idealized, but it was a clear direction and one he was undertaking with typical, quiet resolve.

THE LAST DAY OF summer was September 22. The weather was already turning, fast. Just after 6:15 a.m., Reggie climbed into the SUV to head to a job in Logan. Like always, he took his Cingular flip phone. After he turned east on Valley View Drive, he made his regular stop at the Sinclair gas station for his one-liter plastic bottle of Pepsi. It had started to rain.

At the same gas station, John Kaiserman was pulling up in his Ford F-250, hauling a trailer. For Kaiserman, forty-one, a stout man with a handlebar mustache, the trailer was a kind of mobile office or workshop. He was a farrier, a certified maker of horseshoes, and his trailer carried all the tools of his trade, including nearly one thousand pounds of horseshoes, a gas forge, and a 150-pound anvil. As farriers had done since the Old West, he would visit your farm, assess your horse's hoof needs, "get a piece of aluminum or steel, or whatever your horse required, and I could build it on site, and nail it on." Not bad for the price of $65 to $150 per horse.

The equipment was a hell of a lot of weight to carry around, maybe 4,500 pounds worth. Hence the powerful Ford, itself weighing around 6,000 pounds; together with the trailer, it was nearly five tons—a missile at highway speeds.

For Kaiserman, that morning had been a particularly pleasant one; weirdly so, he thought. He'd awakened naturally thirty minutes earlier than usual. It gave him more than enough time to pitch the hay and tend to the animals on his own modest property, located on five acres just outside Tremonton.

He was comfortably on schedule when he pulled his big load out of the Sinclair station and back onto the road, heading toward Logan. He turned the radio to 96.7, country music. He looked up to see a few snowflakes, and, about two hundred yards ahead, Reggie's white Chevy. It was dark, but Kaiserman was able to see the vehicle wander several times across the yellow divide, then steer back. A few miles later, the Chevy did it again. Kaiserman thought it odd, and he kept his distance. There was no hurry to get to Logan, he thought, no need to tailgate, and the weather was bad. *I got all the time in the world.*

As they pressed on over the curves and hills, Kaiserman saw something that gave him greater pause. The Chevy veered entirely into the incoming lane before recovering with a quick jerk of the wheel. *What was going on?* Kaiserman wondered whether the driver was unsure of where he was going. Or maybe the driver was thinking of taking a left turn on one of the dirt side roads but was having trouble in the low light figuring out which was the correct road.

Or, Kaiserman thought, maybe the driver of the Tahoe was trying to pass the semi just ahead of him. The Chevy, as he put it later, was "half near tailgating the semi." Strange behavior, bordering on very dangerous; why try to pass a semi in the freezing rain?

This guy is an idiot, Kaiserman thought. *This guy is going to cause us all some trouble.*

ABOUT FIFTEEN MILES AWAY, heading out of Logan, in the opposite direction, was a blue 1999 Saturn sedan. Its driver was James Furfaro, thirty-eight, who'd left home that morning a bit late. As usual, he picked up his friend and colleague, Keith O'Dell, fifty, at a Park & Ride in Logan. Both men were scientists commuting to their jobs at ATK Systems, where they were helping build rocket boosters. Rocket scientists. As they drove, Jim munched Cheerios from a plastic baggie handed to him by his wife.

Keith ate his regular breakfast, a red Fuji apple. He was tired, which was his lot in life. He was a contented workaholic. But he'd seemed particularly tired of late, to the point that his wife had suggested that morning that he skip work and stay home.

Around 6:40 a.m., just minutes before dawn, Keith and Jim neared milepost 106.6, which was right around the turnoff to a gun range. Traffic was modest. KVNU, a local radio station, reported the temperature at thirty-three degrees. The roads were wet but not icy.

In the darkness, Keith and Jim could make out oncoming headlights, but not the big, heavy trucks the lights belonged to. First was the semi. Then came Reggie in the Chevy, but he was tracking so close to the semi that he was basically hidden from Jim and Keith's view. Then, a bit farther back, still cautious, drove Kaiserman and his haul. Two minutes earlier, this trio of trucks had sped down the last big hill before Logan—a crest that in the light afforded a spectacular view of Cache Valley below—and they'd descended into a flat patch. It was a straightaway, a relatively easy stretch, though narrow. Reggie felt he knew the road like the back of his hand. He'd driven this hundreds of times to go to Logan—the region's big city—to go to work, see movies, go on dates, hang out, attend all three of his brothers' weddings; he'd taken his driver's test there.

Kaiserman noticed a trickle of snowflakes, not a flurry, intermittent. Then he saw the Chevy slip left again, almost lazily drifting from behind the semi. This time, there was a car coming in the other direction. Even at fifty-five miles an hour, time played a trick on Kaiserman. He had a moment of clarity, not slow-motion exactly, but he could see it coming, something horrible. The Chevy was not returning to its side of the road. Its left front edge was fully crossed over the line. Still barreling at highway speeds, it closed in on the smaller car coming the other way, the distance narrowing by the instant. Things suddenly began to speed up.

The Chevy clipped the side of the Saturn on the driver's-side door. The Chevy bounced off. The sedan carrying Jim and Keith fishtailed. It turned fully sideways. It crossed the yellow divider. Out of control. And then it was right in Kaiserman's path. *Oh shit,* thought the farrier.

He slammed on the brakes. He spun the wheel to the right. It might send him into the ditch, he realized, but at least he'd avoid hitting the sedan square on. It was too late. *Holy shit,* the farrier thought, *this is going to hurt.*

On impact, he heard a crunching sound, somehow modest, a noise that reminded him more of a fender bender than what this was: a high-speed direct hit. The airbag exploded in Kaiserman's face. Foot pressed on the brake, he screeched to the right and saw the front hood of his truck crumple and rise. He skidded to a stop, realizing his driver's-side door was open. Something hurt, his back maybe; it didn't fully register. He climbed out and saw that his Ford had practically bisected the sedan, severed it. It was wrapped around his truck. "My bumper was against the driver's shoulder."

He pulled out his phone and dialed 911.

At the same moment, Reggie climbed out of the Tahoe. He was about a hundred yards away, having finally come to a stop after glancing off the Saturn and then righting himself down the road, his truck virtually unharmed. He saw the wreckage and also dialed 911. But the call didn't go through.

Kaiserman's call did connect successfully. The dispatcher picked up at 6:48:45 a.m., according to the official recording.

"Hurry, send an ambulance."

The dispatcher asked what happened.

"A guy hit a car and it spun him in front of me, and I T-boned him. I think he's dead."

At this point, Kaiserman didn't have a vantage point to see there were two people in the Saturn. The dispatcher asked how many cars were involved, and the location of the accident.

"He's in bad shape. He's bad." He reached inside the sedan to inspect Jim. "No movement, no pulse."

All of a sudden Kaiserman realized there were people everywhere. Sounds, sirens, lights. It was just shy of 6:52 a.m., minutes after dawn, when an emergency medical technician took the phone from Kaiserman.

"We have two 10-85 Echoes here," the EMT told the dispatcher. *Echo* meaning "fatality."

Dispatcher: "*Two* 10-85 Echo?"

The dispatcher asked whether it was possible to start CPR. No, the med tech reported, there was no way to get into the vehicle. The men would need extraction, and they were dead.

Dispatcher: "Is there anybody else injured on the scene?"

No, the EMT said. "Apparently there is a third vehicle that was involved. There are no injuries on that one also." Then he added: "We'll definitely need law enforcement."

TEN MILES AWAY AS the crow flies, the radio beeped three times in the Crown Victoria of Utah State Trooper Bart Rindlisbacher. Just after 6:48 a.m., when the 911 call first came in, the dispatcher called out a 10-50 PI, a personal injury accident, and the coordinates.

Trooper Rindlisbacher pulled a quick turn and headed for the scene. He'd started his shift at six a.m. and was still getting his bearings. Rindlisbacher was on his first week back on the law enforcement job after doing a tour in Iraq. He'd done security convoy, escorting trucks to the Turkish border, dressed in full body protection, hoping neither he nor the trucks would stumble onto land mines or IEDs.

Even before his tour, he was no stranger to carnage. He'd worked in the army as an emergency room tech in Korea and at Fort Lewis in Washington. He got used to shutting out the noise and danger; at night, on the base in Iraq, he'd pull on eyeshades and his headphones and listen to classic rock. Once, he'd slept through a mortar attack.

He managed in Iraq to also train for a marathon that was a mere few weeks away. Rindlisbacher was a man with a reputation for tenacity.

A local Logan city police officer came over Rindlisbacher's radio. The cop had arrived at the scene. "Possible Echo," the officer reported. Another flurry of snow.

Minutes later, Rindlisbacher arrived at the scene. There were already fire trucks. The trooper looked inside the Saturn. No doubt: Echo. The bodies were tossed and charred. Both men appeared to have been killed on impact.

Because Rindlisbacher had only been back on the job for a week, his car hadn't been fully equipped. He pulled out his own personal camera and began shooting pictures before anything got moved. A decidedly nasty wreck, he concluded. A collision so violent it popped out the passenger's eyeballs.

Rindlisbacher started looking for witnesses; the local cop showed him to Kaiserman, who was sitting in the back of an ambulance and gave the trooper his account of the swerving Chevy Tahoe. The cop also pointed Rindlisbacher one hundred yards down the road to a young man standing beside a white sport-utility vehicle.

Rindlisbacher walked down to Reggie. It was now clear he'd been the driver of the car that clipped the Saturn. The trooper made a quick assessment of the young man. Six feet tall, around 150 pounds, quiet. He also discovered that Reggie's mother was there. She'd arrived soon after receiving a call from Reggie. Even before he started talking to Reggie and his mom, Mary Jane struck Rindlisbacher as a take-charge kind of person.

As he got his bearings, he glanced at a statement Reggie had already written out for the local police. The first officer on the scene, Chad Vernon, who arrived minutes before Rindlisbacher, had asked Reggie what happened. The young man told Vernon that his car had hydroplaned.

"I was driving east toward Logan," Reggie then wrote in a statement with neat-enough letters that lean slightly left. "My car pulled to the left and I met another car in the middle. We clipped each other, and he spun out behind me. The truck and trailer behind me then T-boned him and they ended in a ditch."

Rindlisbacher introduced himself and asked if Reggie would be okay with taking a drug test. Sure, Reggie said.

Trooper Rindlisbacher said he needed to take Reggie to Logan Regional Hospital. Reggie's mother offered to do it instead. Rindlisbacher felt she was pressuring him. "I'm going to have to ride with him and ask him a few questions," he told her.

Just before eight a.m., Reggie climbed into the passenger seat of the Crown Vic. He didn't say anything.

"You understand two guys died today," the trooper said. Reggie nodded his acknowledgment.

Rindlisbacher probed gently into what might've caused him to come across the median. Poked around, looking for an explanation. Reggie reiterated that he thought he'd hydroplaned. End of conversation.

But that answer kept nagging at the trooper. Reggie's SUV must have weighed four thousand pounds. It wouldn't hydroplane unless it was going one hundred miles an hour. The witness, Kaiserman, said they were going fifty-five miles an hour, the speed limit. Plus, of no small significance, Kaiserman said he'd seen the Chevy swerve several times prior to the crash.

A few seconds later, from the corner of his eye, Rindlisbacher saw Reggie reach into his jacket. The young man pulled out his phone. The ringer was on silent, but Rindlisbacher could see Reggie had gotten a text. Reggie replied to the message, and stuck the phone back in his pocket.

Over the next few miles, Reggie did it four or five times. Something about it struck the trooper. "He did it with one hand, held the phone and texted with his thumb. He was a one-hander."

CHAPTER 2

THE TOLL

LEILA O'DELL NEVER WORRIED about her husband, Keith. Why bother? He was good at everything. Carpentry, plumbing, tree pruning, and computers—especially computers. The basement in their North Logan home was an electronics graveyard that included Keith's first Atari, an ancient model he'd sometimes pull out, disassemble, then solder back together as an improved version. His ATK Systems colleagues would come to him for computer advice rather than go to the IT department. Even among those powerful engineering minds, he stood out. His nickname was The Genius.

ATK built rockets under a contract from NASA, which in 1999 awarded Keith the Silver Snoopy Award, a prestigious honor reserved for people who show great dedication in bringing safety to flight missions.

But as capable as he was, Keith had an even bigger sense of duty. He was the man in the background who got things done. At his daughter Megan's swim meets, he ran the computer that kept time and the score. And, at home, he was the guy with the tools.

Leila thought of him as a do-everything guy. "A woman once asked me: 'How do you get your husband to fix the plumbing?' I said: 'I just tell him the sink drips, and he takes care of it.'"

One thing Keith didn't do all that well was talk. He and Leila had met in high school in nearby Box Elder County, and ever since, they had walked, biked, and read together, and enjoyed each other's company, often with few words exchanged. He brought out her joie de vivre, a passion, an almost infectious laugh.

Keith had been the product of a Protestant and a Catholic. Leila was Mormon. But they didn't need what she thought of as the "social club" of church. They had each other. Politically, they were conservative, closer to libertarian than liberal.

"We were just quiet people," Leila sums it up. "It was never fireworks—just quiet, calm, efficient, responsible—and that just suited him."

Leila worked part-time as a bookkeeper with a heightened attention to detail; her memory for little things bordered on the photographic. Friday, September 22 wasn't a workday. She had planned to work in the yard, only to look outside to discover snow.

She'd actually hoped Keith wouldn't go to work that day. He'd been working so hard, dealing with a new contract. The night before, he hadn't come home until 8:30, and went right down to his computer in the basement, at the desk strewn with computer discs, and started working again. She brought him leftover sweet-and-sour pork, and went upstairs and read until he joined her.

Of course, that Friday, he didn't heed Leila's suggestion to take the day off. He worked; that's what he did.

AT ABOUT 9:40 A.M., Leila was still wearing her pajamas covered by a floor-length purple robe when the doorbell rang.

She opened the door and discovered a local police officer and a sheriff. She immediately thought something must have happened to eighteen-year-old Megan, the girl Leila and Keith had adopted, taking her home from the hospital on Friday the 13th of May in 1988. Megan had been a daddy's girl. But starting in high school, she struggled with grades, turned away from her promising swimming talent, and, just generally, caused Leila ceaseless worry. The men stood with Leila just inside the doorway. They could hear the wind playing the chimes that hung just outside the front door.

The law enforcement team explained to Leila that there had been an accident. She braced herself. The wreck involved a Saturn belonging to Jim Furfaro. There was a passenger, too, who they suspected could be Keith, but, well, the crash had been intense.

"They couldn't identify him," Leila says.

There must have been some mistake. Surely, Keith, the genius, the man who never ran afoul of anything, couldn't possibly have been the passenger in a fatal wreck.

MINUTES EARLIER, THE POLICE had pulled into Utah State University, the workplace of Jim Furfaro's wife, Jackie, a computer programmer. She had gotten to campus a little after nine, after having dropped off the couple's two daughters: three-year-old Cassidy at day care, and Stephanie, who had just turned seven, at school.

Jackie had a cold that morning. In fact, she, too, had thought about asking her husband to stay home from work to help with the girls but decided otherwise. Before he'd left that morning to pick up Keith, a few minutes late, the couple had discussed what they might do over the weekend. Jim thought he might attend a tai chi retreat. If not, given Jackie's cold, the pair thought they might play World of Warcraft, a strategy computer game. They would often play together, side by side, at computers in their basement. Near the computers was a television on which Jim and his daughters played Dance Dance Revolution on the Nintendo Wii.

But Jim, who had studied mechanical engineering with an aerospace emphasis, wasn't a computer geek. He was game for trying everything at least once. He rode a unicycle. He painted the cartoonish, alien-looking black creature on the wall over the kitchen table that the family called the "funny man." Jim was comfortable with himself. He made other people feel comfortable.

As Jackie hustled down the hallway to her office that morning, an administrator asked if she might come into an empty classroom. Once there, she saw the police officers and a woman from her neighborhood. Her first thought was that she must have caused an accident.

The police asked her to sit down and explained there had been a wreck on Valley View Drive. She wasn't getting it. They told her that the incident had taken place on the road to Tremonton.

She was trying to make sense of it when she noticed that one of the policemen had something on the desk in front of him. She looked at it. It was Jim's driver's license.

"No, no, no, no, no!"

When she got her bearings, they asked her who else might've been in the car. She told them: maybe Keith. Then she asked them to tell her again what had happened.

"Jim's car was clipped by a Tahoe that had crossed the yellow line and sent it into oncoming traffic," an officer told her.

"That was pretty much it," she says. There was no further explanation. What can ever explain this kind of tragedy?

NEARBY, AT LOGAN REGIONAL Hospital, Trooper Rindlisbacher was also having trouble making sense of the events. He was sitting in his parked car in front of the emergency room, asking Reggie questions, this time to fill out the paperwork. He was eliciting a lot of "monosyllabic" answers.

Was Reggie tired? No, he'd had a good night's sleep.

Was he on any prescription medications? No.

Were his wipers working? Yes.

Was his defroster working? Yes.

Rindlisbacher was typing the answers into a standardized form on a laptop attached to a metal platform extending from the Crown Vic's center console. By 2006, police cars had begun to get more and more high-tech. There was a camera on the front of his car that fed video to a VCR in the trunk. In the backseat, there was a printer. It was hooked up to the laptop, a Panasonic Toughbook, which allowed the trooper to run license plates and do paperwork.

Curiously, given how high-tech the cars had become, and the preponderance of cell phones, not a single state at that time banned texting and driving. In fact, most jurisdictions in the country had no mechanism for reporting whether a driver had been on a phone, even talking, let alone texting. In Utah, there were no laws prohibiting either activity, except among novice drivers.

Rindlisbacher said he had a final question.

"Were you talking or texting on the phone when you were driving?"

"No."

"I don't understand, you say you were hydroplaning." It was a statement but meant as a probe, an effort to get Reggie to explain himself.

Yep, Reggie nodded, hydroplaning, that's what happened.

RINDLISBACHER, THROUGH YEARS OF experience, knew that sometimes there were accidents, but more often, there were wrecks. Things happened for a reason. On the roads, that reason was usually booze. Of the 43,500 fatalities on U.S. roadways the previous year, 17,590 had been caused by drivers with a blood alcohol level of .08 or higher.

Most of the fatal crashes involving just one car, like a guy driving into a tree, happened at night. But the multiple-vehicle wrecks tended to happen during the day, which stood to reason, given that there are more cars on the road at that time.

In Utah, as elsewhere, teen drivers caused a disproportionate number of crashes. Statewide, teen drivers caused 26.8 percent of all crashes in 2005, and they caused 31.8 percent in Cache County, where that morning's incident had taken place. Of the six deaths that occurred in wrecks the year before in the county, one was attributable to a teen driver.

How many crashes nationally were caused by cell phones was not yet clear, but the emerging evidence was alarming. In 2003, researchers at Harvard University did a risk analysis and projected that motorists distracted by their cell phones caused 2,300 deaths each year and 330,000 injury accidents.

That research was based largely on people dialing and talking on phones, holding them to their ears.

But the opportunity for risk was growing, given the exploding power of devices. Text messages had been sent earlier than 1999, but that was a key year because a Japanese phone carrier, NTT DoCoMo, built the i-mode networking standard, which allowed for the exchange of mobile data. By 2002, according to the Computer History Museum in Silicon Valley, more than 34 million subscribers were using that network for web access, email, and other functions.

And 2006 was an important year for the mobile phone because of the growth of smartphone devices. An article in *BusinessWeek* billed 2006 "the year of the converged device." It reported that 80 million smartphones had been sold. They were "smart" because they could do so many things beyond calling, like texting, emailing, surfing the Internet, playing games. "Some phones are now below $200, encouraging the clever phones to spread beyond corporate users and into early adopters in the consumer market," the article pointed out.

Most people didn't yet have a smartphone. Reggie didn't. But a smartphone wasn't needed to send simple texts. Those you could do with the older phones, so-called feature phones. By 2006, people in the United States were sending around 12.5 billion texts a month, which sounds like a lot, but the technology really was embryonic. Two years later, there would be 75 billion texts sent a month, with around half of them sent by people under the age of thirty-five. Like a lot of new technology, it skewed toward the young. Rindlisbacher was among those with only a passing knowledge of texting.

But he sensed something was amiss.

WHAT IRKED TROOPER RINDLISBACHER was Reggie's written statement. *My car pulled to the left and I met another car in the middle.*

To Rindlisbacher, this made it sound like Reggie was "rubbing off some of the blame." *We met in the middle.* To Rindlisbacher, it was like Reggie was trying to justify it, "to make it sound like it was partially the other guy's fault."

What Rindlisbacher didn't know was that there were reasons why Reggie wasn't saying much. One was that he was genuinely having trouble remembering what had happened. It had all unfolded so quickly.

The other explanation had to do with a phone call. It was placed less than an hour after the accident. Reggie's mother, moments after she'd arrived at the scene, had dialed her oldest son, Phill.

When the call came in, Phill, coincidentally, was himself driving. He was in West Sacramento, California, a lawyer heading to his job at a firm that handled property-loss cases for State Farm insurance. Phill felt compelled to answer because he was surprised to hear from his mom at that time of day.

When she told him what had happened, Phill's mind split in two—"big-brother mode and attorney mode," as he put it later. He had no idea what the facts were, other than that two men had died. He was imagining the civil liability, for Reggie, and his folks.

The Shaw family was close, kin, but Phill and Reggie were separated by eleven years. The Reggie who Phill knew was a friendly little guy, a competitor on the football field and basketball court, as well as with the video game controller. His kid brother loved those video games. He also thought of Reggie as the son who tended to do whatever his mom and dad asked with no protest, the kid who cleaned up after himself.

Now he'd been involved in an accident with two fatalities.

What, their mother asked Phill, should Reggie do?

"Don't say anything to anybody."

When they had arrived at the hospital, Mary Jane passed along word to Reggie. "Don't say too much," Reggie says his mother told him. Leave it to the police to piece it together. As she puts it, she instructed him: "Let them figure things out, and we'll go from there."

ABOUT AN HOUR LATER, sitting outside the emergency room, State Trooper Rindlisbacher gave up on getting answers. He escorted Reggie inside, where the young man took and passed a blood test. And it seemed there was little for the trooper to do. He could've written a "left of center" ticket, a moving violation that would have laid the matter to rest. Plenty of law enforcement personnel would've done exactly that and not been unjustified in so doing.

"Sometimes I like to follow things through and follow up," Rindlisbacher reflects of himself, adding with a laugh, "Some of my coworkers think I'm too thorough."

Once in a while, his tenacity invited citizen complaints. There was the time he got investigated by internal affairs after being accused by a woman of stalking her family. Rindlisbacher was exonerated. He wasn't stalking, he says; he was trying to ferret out a liar in a set of bizarre circumstances. It had started very innocently when he pulled a guy over because a Christmas tree tied to the roof of the man's car was falling off.

Then, coincidentally, Rindlisbacher caught the same guy speeding a few weeks later, this time driving a different car. What piqued Rindlisbacher's interest was that both of the cars had Idaho plates, and the guy had an Idaho driver's license. The man insisted he lived in Idaho, even though Rindlisbacher had pulled him over twice locally, in Logan. The trooper did a little background check and discovered the guy had a prior arrest for meth possession. It's a tax-related crime in Utah to reside there over an extended period of time while claiming you live in another state.

But the guy kept claiming he lived in Idaho. It didn't add up.

So Rindlisbacher says he decided to find out for himself. Every few weeks, he drove by the family's residence and took pictures, to prove they lived in Utah, not Idaho. It turned out, Rindlisbacher explains, that the wife of the guy was the daughter of a police captain in Las Vegas. The captain suggested she call Rindlisbacher's headquarters and file a complaint. She did—for stalking.

"People complain about the stupidest things. They don't want to take responsibility for their actions," Rindlisbacher says. It really ticks him off when they don't. "That's why I say: 'Tell me the truth and there's no problem.'"

Rindlisbacher was sure Reggie wasn't being square with him. But how to get at the truth? Maybe he could talk to Reggie again. After a few days' reflection, Rindlisbacher hoped, the young man might explain better why he'd wandered across the yellow line—and more than once.

CHAPTER 3

THE NEUROSCIENTISTS

T HIS IS MICKEY HART'S brain."

Mickey Hart was the drummer for the Grateful Dead. An image of his brain appears on a twenty-four-inch computer monitor. To its right, another monitor, a sleek thirty-two-inch Mac, features a splash of windows—email, news sites, a work project.

Mickey's brain is red on the top with blue sticking out on the bottom. "The red is the cortex."

Few know more about Mickey Hart's brain, about the brain in general, than the man pointing to the monitor. His name is Adam Gazzaley. He's a neurologist, an MD, with a PhD in neuroscience. He runs the new neuroimaging lab at the University of California at San Francisco, one of the world's leading scientific institutes.

Dr. Gazzaley's lab is housed inside the Sandler Neurosciences Center, a five-story, 227,000-square-foot research facility that opened in May 2012. Located minutes from downtown San Francisco, a baseball's throw from where the San Francisco Giants play, it is a gleaming example of a new dedication to understanding the workings of the human brain. That pursuit itself is nothing novel, of course, but now, a new generation of powerful technology lets researchers see the inside of the brain, watch it work, literally, and observe when it fails to work.

Dr. Gazzaley's lab contains around $10 million worth of equipment that the researchers speak of only by acronyms, the fMRI (functional magnetic resonance imaging), the EEG (electroencephalography), and the TMS (transcranial magnetic stimulation). With the machines, the scientists study blood flow in the brain, look at electrical wave patterns, and create images of ultra-thin slices of neurological tissue. The various techniques let researchers understand which brain regions control what functions, and how tissues and tasks get impacted by different activities, say, when a person tries to multitask.

Mickey Hart's is one among many brains Dr. Gazzaley has imaged. The pair have been working on a pop science project together in which Dr. Gazzaley shows what Mickey's brain looks like while he's drumming, trying to elucidate not just the brain of a rock star but that of an aging one. They do presentations where Mickey drums and Adam shows off images of the percussionist's brain taken in real time using sensors attached to Mickey's head. Your Brain on Rhythm.

Dr. Gazzaley himself might pass for a hipster musician. He's a youthful-looking forty-five, with short-cropped silver hair—not gray but silver—that looks like it's been dyed to get attention, even though it's been the same color since it prematurely aged in his early thirties. He wears a serpentine ring on his right index finger. He tends to sport black jeans that are on the tight side, and a silk shirt. His car is a BMW M3 convertible, the super-fast kind. He's become friends not just with Mickey but also with the lead singer of Thievery Corporation, a rock band, as well as some of the tech billionaires who attend the late-night parties he holds on the first Friday of each month.

A few months earlier, Dr. Gazzaley had gone to Germany to speak at a conference. At the airport in Berlin, a woman at immigration control asked him his business. He explained that he's a scientist.

"Really?" she asked.

"Yes."

"You don't look like one."

DR. GAZZALEY TRAVELS A lot. He'll put on 150,000 miles a year on airplanes, give or take. He gives upward of fifty talks.

"Sometimes I ask myself: 'Why do I constantly put myself in these stressful situations?'" he says. "It's not like I have to do it."

The toll that Dr. Gazzaley is referring to comes in large part from the challenge of juggling all of his responsibilities. The thirteen people he supervises, the constant fund-raising, the media appearances. He regularly forgets where he parked his BMW in the adjoining parking structure because he's so busy thinking of other things when he gets out of the car and walks into work. Or he's fiddling around on his iPhone. Once, lost in thought while brushing his teeth, he put moisturizer on his toothbrush, not toothpaste.

Dr. Gazzaley isn't particularly absentminded. He simply feels like he's experiencing a pressure so many people feel in their everyday lives—to keep up, push on, achieve.

"Everyone feels that burden in their own way of trying to squeeze as much out of our brains per unit time as possible."

We're all struggling to maximize our attention.

Attention.

Dr. Gazzaley is one of the world's foremost experts in the science of attention. He's consumed with how we focus, what causes us to lose focus, get distracted. The paper that first brought Dr. Gazzaley his notoriety, published in 2005, showed the key parts of the brain circuitry involved when a person ignores something, or tries to ignore something. That science of ignoring is a key part of the attention conversation. Can we ignore what we want to, even need to, in order to survive? One experiment after another of his explores how we focus, what takes us away from what we profess to want to focus on, and whether our attention limits might even be expanded. He's spent four years on an experiment looking at whether a particular kind of scientifically engineered video game could improve the attention span and memory of people over the age of sixty. He's been on pins and needles lately, hoping the results of the experiment could land in *Nature*, one of the leading scientific journals in the world.

And he's trying to distill all of these ideas into easily digestible chunks—science meets pop science. For instance, when he first showed off Mickey Hart's brain, during an AARP event in New Orleans, Dr. Gazzaley kept bringing the conversation back to attention and distraction.

Wearing a baggy shirt and orange-tinted glasses, Hart was fitted with a wireless sensor on his head that fed his brain waves into a computer. The signal was then transmitted to two giant screens on the stage.

For his part, Dr. Gazzaley spoke into a wireless mic and paced, holding an iPad to control what images of Mickey's brain appeared on the screens.

"These are the theta waves," he told the audience. "These are associated with attention and concentration."

Back in his San Francisco office, Dr. Gazzaley pulls out a white plastic model of a brain. He pulls the hemispheres apart. He holds the left half in his left palm. He runs his other hand across the outer part, the wrinkle near the front, the red place he'd described earlier in the virtual version of Mickey Hart's cortex.

It's the proverbial gray matter, the most evolved part, he explains, the part that makes us human. "It's where the action is." Abstraction, language, how we make decisions, organize our time, focus.

He moves his finger lower down on the brain, toward the stem. This, he explains, is the figurative and literal lower region. The reptilian part of the brain that resides in most animals, highly evolved primates and otherwise. It told our forebears whether to run when they saw a lion or perk up at the sound of a bird that might be food.

"It controls quick reactions to things. It deals with basic stuff, like seeking mates and reproduction. It's been preserved throughout evolution."

Among Dr. Gazzaley's many research areas, he has explored the tension between these relatively primitive parts and the more evolved regions, including the frontal lobe. How we balance the tension between the short-term demands of our reptilian senses—run!—and our longer-term desires, goals, and commitments that we try to set with the more evolved parts of our brains.

It's why I've come to visit him on this gray day in early December 2012. I'm trying to better understand why two rocket scientists are dead. Was it because Reggie, for some reason, lost his focus? Was he distracted? What was happening inside his brain? Can the research being done here, and by a new generation of neuroscientists, prevent similar tragedy? Does science offer any solace, some hope, for the families of Keith and Jim?

More basically: What is attention?

DR. GAZZALEY SMILES.

"What is attention? Attention is a complicated thing." He pauses. "It has many subdomains. It falls into an aspect of cognition that's related to the selection of information to be processed."

Oft cited on this subject of attention is William James, a philosopher and physician at the end of the nineteenth century, and the brother of novelist Henry James. William wrote of attention in 1890: "Everyone knows what attention is. It is the taking possession of the mind in clear and vivid form out of what seem several simultaneous objects or trains of thought."

Everyone might know what it is. *Might*. But, as Dr. Gazzaley notes, more than a century after Dr. James, what we think we know, or what we once thought, has many strata. It's complicated.

Dr. Gazzaley says that attention is also "absolutely critical for all high-level functioning," a cornerstone of what it means to be human. He's not just saying that our attention allows us to survive—say, by being able to attend to a threat or perceive an opportunity. He means that attention allows us, in a "uniquely human" way, to set goals and follow through on them without being distracted by every bit of stimulation around us.

"It allows us to interact with the world through our goals and not be led by or be a slave to our environment. It has allowed us to do every remarkable achievement—creation of society, culture, language. They are all dependent on being able to focus on our goals."

To illustrate how attention works, Dr. Gazzaley has an idea. He suggests I come to his next First Friday cocktail party, his monthly hipster soiree.

Learn about attention at a party?

"I'll show you the cocktail party effect."

CHAPTER 4

REGGIE

"MOM, WHAT'S GOING TO happen?" Reggie asked.

"Everything's going to be okay, Reg."

Mary Jane's voice could strain a bit when she got excited.

She and Reggie, in particular, had a playful relationship. He'd tease her, gently, but he was always listening and always attentive.

He sat in the passenger seat of his mom's Chevy Blazer. It was just hours after the accident, and Mary Jane was driving him home from the hospital. She'd decided to take an alternate route, using side roads, circling around a chunk of Valley View Highway, so Reggie wouldn't have to pass the site of the wreck, where mop-up and accident reconstruction were still under way.

They hit the crest and started sloping back down toward Tremonton. They passed through Beaver Dam, a small town. It was light now, raining.

Waiting at home was Reggie's dad, Ed. He wasn't sick with worry yet. He'd been heading out the door to work when Reggie had first called from the side of the road. Initially, he had just said he'd been in an accident, and Mary Jane had told Ed it didn't sound that bad. Reggie said someone had swerved into his lane and that he was unhurt. So Mary

Jane sent Ed off to work at the machine shop where he was the foreman.

But as soon as she got to the scene and realized what had happened—that two men were dead—she called Ed, and he went right home to wait. There still wasn't any indication Reggie could be at fault, but Ed obviously wanted to be there for his son.

"What's going to happen?" Reggie again asked his mother. "I don't want to go to jail."

EVERYTHING WAS GOING TO be okay, she told herself. Reggie was a polite boy, and the son who'd already survived a few challenges—the concussions in football; that goofy, lovable awkwardness; the odd birth. The day he was born, in late January 1987, Mary Jane's doctor wanted to be there for the birth, but also to leave on a family vacation. He induced labor by stripping Mary Jane's membranes at his office and, several hours later, Ed hustled her to the hospital. The baby came a few hours later, on the early morning of January 28. They had to suction his head to get him out.

"He came out with a cone head." Mary Jane laughs when she tells the story. "I told the doctor to put him back in because he wasn't done yet."

She and Ed hadn't yet thought of a name. A black-and-white television hung in the corner of the delivery room showing Pope John Paul II. His name, in Italian, was Giovanni Paolo II. It sounded a bit like Reggie.

"It was kind of sacrilegious," the devout Mormon says of naming him after a pope. "But it just kind of fit."

Reggie had been the second-to-last child of six, the siblings spanning in age from Vicki to Phill to Jake to Nick, who was eighteen months Reggie's senior and his pal and constant competitor. Later came Whitney, born in 1999, when Reggie was almost thirteen. The Chevy they drove was nothing fancy, a standard model. But "standard" by then included some remarkable safety technology, like an antilock braking system, which allowed for swifter stoppage time and had first been developed in the 1960s. It had power steering. These were technologies

that had been part of a profound effort to make cars and roads safer—
along with airbags and billions spent to widen and improve roads. And
yet traffic fatalities continued to be a tenacious problem.

The car didn't have a navigation system. Those were in their infancy,
with the first TomTom GO GPS first appearing in the Netherlands in
2005. Soon they were everywhere. That's how fast things were moving
on the technology front.

Mother and son drove home from the hospital mostly in silence.
When they got to the red-brick house, Reggie marched right up to his
room. He lay down on the twin bed that was pressed against the far wall
and turned on his side, away from the door. He put his cell phone on the
bed, behind him. He was exhausted, terrified, and unable to find any rest
or peace. At the hospital, Rindlisbacher had suspected that might happen.

"Reggie's probably going to have trouble sleeping. Can you get him
something to help, if he needs it?"

This room was a particularly tough place to let his head clear. Not
because it wasn't quiet, but precisely because it was so quiet. Hardly
any traffic, not even birds, not at this time of the day, in the drizzle.
Through the window next to his bed, he could see the wooden jungle
gym with the green slide, next to a one-bedroom brick cottage in the
family's backyard that had in years past been home to local missionaries.

Reggie tossed and turned. He looked at the ceiling, the red trim on
the closet door, the bullhorn that hung over his bed, the radio station
sticker on his door. His phone rang. He turned over and, without looking at it, sent the call to voice mail. It rang again. And then it started
ringing regularly. Word was spreading. Reggie prayed. *Please let me go
back and do it over. I would trade anything to go back.*

HE TRIED TO FIGURE out what had happened, to piece it together. The
last thing he remembered before the wreck was being on the opposite
side of the road. But he couldn't figure out if he'd been the one to
drive there, or if something had hit him or someone had pushed him

there. Try as he might, he couldn't remember any sound of cars colliding.

His first vivid memory afterward came as he walked from his truck toward the Saturn. There was this stout man with a handlebar mustache who was getting off the phone with 911.

"What happened?" Reggie asked.

The man was Kaiserman, the farrier. He responded: "You just hit that car."

TOM HIGGS, A MANAGER of advanced engineering at ATK Launch Systems, saw Brian Allen stand and walk to the center of the room, right in the middle of a dozen cream-colored cubicles. Brian looked grave. "That wreck—I think Keith and Jim were involved," Brian said. He was one of Keith's best friends.

Tom and the rest in the group had wondered why Keith was late, and they'd heard separately that there had been an accident on Valley View Highway. It was the route many at ATK took to this isolated desert spot situated roughly between Tremonton and Brigham City. You had to be in the middle of nowhere when you were building and testing rocket boosters—not just for the space shuttle, but for military missiles, too.

Brian had no further information, and returned to the phone to try to find out what happened. His seat was located near Keith's in the mostly uniform batch of cubicles. But Keith's work area stood out, for two reasons: He had a bigger cubicle, reflecting his status as a senior analyst and lead engineer, and his space appeared to be a mess. It was packed with piles of papers, journals, and reference materials, each standing two or three feet high. Stacks on the floor, stacks on the cabinets. To get Keith's counsel, you kind of had to sidestep your way into his cubicle to avoid the papers.

And that's what people did: came to get Keith's advice. They'd do so for engineering, of course. But they'd seek his help for physics and even software coding, a tribute to a rare depth and breadth of scientific capability. Over in the production group at ATK, a whole different division, they sometimes had trouble with a machine that inspected the booster rockets' O-ring grooves to make sure there was no corrosion. The production guys actually would turn to Keith to fix the software.

His chief charge, though, was designing the next generation of rockets. When the government or a defense contractor put something out for bid, the engineering team would use their computers to simulate new designs; the idea was to push the limits of physics and math, and of known resources and production capabilities, to create new generations of powerful, stable rockets.

There were major-league stakes: intellectual and scientific, financial, and, less often discussed, military. ATK and its denizens might've bragged about their relationship with NASA and the space shuttle, and they were absolutely integral in helping redesign its boosters. But the engineers also spent time designing strategic rockets under defense contracts, something perhaps less politically correct to advertise: Trident missiles, ICBMs, Minutemen. Nuclear weapons. The most powerful and deadly things in the world.

Until several months earlier, Jim had worked in the same bunch of cubicles as Keith. Then he moved to another part of the building to take over a managerial position on a key project: developing a new, more powerful booster for the space shuttle.

Brian, Keith's close pal and Jim's friend and colleague, returned to his cubicle to take a phone call and, less than a minute later, returned to the open area with an update. He was practically matter-of-fact about it—a reflection of disbelief or, perhaps, the communication style of the engineering culture.

Jim and Keith were dead.

VAN PARK STOOD AT the podium, shuffling papers, as the fifth-period students began pouring in. A boy approached.

"Coach Park, did you hear what happened to Reggie Shaw?"

Van shook his head.

"He got into a really bad accident."

It was early afternoon on September 22, a few hours after Reggie's car swerved left of center.

Van wore khakis, a golf shirt, and tennis shoes; he stood six feet tall and weighed 185 pounds, fit, his head a bushel of blond hair. On the wall of his classroom hung two flags for the Utah State University Aggies. There were posters warning students about the dangers of drugs and STDs. And there were three framed nature pictures, each one urging excellence with a word: *Challenge*, *Determination*, and *Success*.

It was room 157 at Bear River High School—coincidentally, the room that Reggie had taken his driver's education class in during his sophomore year.

Van took a deep breath. The longtime health teacher and varsity basketball coach occasionally heard the rumors that whipped through Bear Valley High—so-and-so lost his arm in an accident, or so-and-so got arrested—and he always moderated his reaction until he heard what had actually happened.

Two thoughts went through Van's mind: *I hope Reggie's okay*, and, *Valley View Drive can get real slick when it's wet. It must have been one of those bad roads.*

The boy reporting the news to Van had no further information—was Reggie okay, had anyone been hurt? *Shrug*. Van glanced out the window, into the seminary, a Mormon teaching center where students are allowed to spend a period each school day learning the gospel and earning credits toward seminary graduation.

Van tried to put these thoughts about Reggie out of his head, even as the rumors continued to swirl throughout the day. But Reggie had graduated only eighteen months earlier. And this was a small town in every conceivable way—size, population, culture. They supported one another, helped out in all the small-town ways. There were also dark secrets people didn't much talk about, along with a gray market of gossip that could spread the ill-informed rumor as fast as any IM or text network.

TREMONTON HAD BEEN CLOSE-KNIT since it was settled in 1888 by folks of German descent by way of Tremont, Illinois. The settlers were largely Protestant, as compared to their Mormon neighbors. In 1903, when they formally incorporated as a township, they called their new home Tremont. But the postal service in Utah feared confusion with the town of Fremont. Hence: Tremonton.

Then Tremonton started to receive a heavy dose of Mormons, spreading out from Salt Lake City. They made significant outposts in Brigham City, just twenty minutes from Trementon, at the base of the mountains, and in Logan, the region's hub, which held a larger temple. It was a place for marriages and sacred ordinances, not an everyday churchgoing experience.

Logan was at the base of the mountains, too, the ones that defined Utah in so many ways. When the prophet Brigham Young reached the base of the Wasatch Mountains, he reportedly declared, "This is the place." The mountains proved a natural border of sorts. They helped explain why the community could grow so close, the families in the cities so interdependent, and their ideas in some ways so insular. It wasn't easy to go over the mountains—either coming in or out, bringing ideas or taking them away.

Tremonton lay on the valley floor, making it good for farming. For many years, it had an odor, a pleasant one—sweet and a little moldy. That was the product of the sugar beet plant located two miles north of town, its aroma carried easily even with a light wind. The farms, like the seven-hundred-acre one that Reggie's maternal granddad, Wilford, once tended, grew beets, mostly, and corn and wheat. There were cattle ranchers and dairy farmers, too.

Everyone knew everyone, proverbially, and, in the case of the Shaws, it was probably closer to literally. Mary Jane and Ed had gone to school there, and married at a little reception center on Main Street when she was barely eighteen and he was nineteen. Wilford, Mary Jane's dad, who was known widely as Wit, was gregarious and easygoing, a pillar in every way.

Ed's family had a rougher go. His mother had had a debilitating stroke when Ed was eleven; one of seven children. The four older ones had more or less left home, but that left Ed at home with a mother who had lost her ability to speak and walk, and a father with a serious drinking problem. He wasn't a binge drinker, and not a mean drunk, but he'd be tough to match for consistency. He managed to get himself to work every day, as a driver and odd-job doer for one of the town's wealthy businessmen, and then he'd come home at night and head for the bottle.

His drink was whiskey. He'd hide it in the car. Every night, numerous times, he'd say: "I've just got to go to the car." He thought he was being sneaky, but he fooled no one.

With his eye on the bottle, it wasn't on Ed. The young man would spend many nights away from home with friends; his absence, he said, unnoticed. He smoked cigarettes, and he drank. He stayed in school, mostly to play sports.

Then he met Mary Jane and got married, and she got pregnant. He was determined to be a more attentive father than his father had been to him. He swore off smoking and drinking.

As a family, Mary Jane, Ed, and their six kids had a reputation for kindness and for their love of sports—the kids playing, and everyone watching and rooting. They'd go to the Utah State football games and travel to Las Vegas for the regional basketball conference championships. The TV was always tuned to some game or another. They never missed a game of Reggie's.

While he wasn't a star, Reggie was a good enough athlete to develop his own modest reputation in the sports-crazed community.

Maybe that's why, on the morning of September 22, news spread so quickly that Reggie Shaw had been involved in a bad wreck. Maybe it's because everyone thought that Reggie was such a good guy.

INDEED, FOR VAN, HIS old basketball coach, Reggie stood out because he was, in addition to being a strong guard with the will to rebound, a decidedly decent person. Van and his wife, Lisa, a French teacher at the high school (her classroom is directly above her husband's) privately said how they wished they'd find someone like Reggie for one of their daughters. To Lisa, Reggie was a model kid: friendly, widely liked, quick-witted, but never seeming to crave the limelight. He got B-pluses and A-minuses in French, and, even though he was a jock, he "didn't mind being called on in class." Lisa thought of him as "the all-American boy."

As Van stood there waiting for the fifth-period kids to finish filing in, he flashed on a moment from nearly two Januaries earlier. It was a day that had started with such high hopes: the varsity hoops team busing to Salt Lake for the state tournament. The team had reason for optimism. They were stacked with talent—big, tough players, like Jason Zundel and Dallas Miller, and they had in Reggie a tenacious defender and supporting player. He was the guard who preferred passing to shooting, rebounding to glory, was best friends with the studs, a bit in their shadow; plus, Van thought, Reggie lacked the confidence to equal his talent and hard work.

For all the team's talent, the state tournament did not go well. The team lost a heartbreaker in an early round and returned home the same day. Hours later, after it seemed everyone else had cleared out, Van walked into the locker room and heard the crying.

Reggie sat with his elbows on his knees and his face down. He looked up.

"I can't believe it's over."

"That's why I love you, Reggie. You're so passionate about basketball."

But Van knew Reggie's reaction was about more than a game—

the season-ending loss meant the end of relationships and feeling connected, something the young man seemed to crave. Looking back on that day, Van thought of Reggie: "He was almost as broke as you'd see with someone who lost a loved one."

On the day of the accident, Van couldn't get the idea out of his mind that something bad had happened to Reggie. When he got home after school, he called the Shaws. He got Mary Jane. She said Reggie was okay. She said there had been "fatalities." It was a word that, for some reason, Van picked up on. She didn't say "dead," she said "fatalities."

Reggie was upstairs. He'd not come down all day. Hadn't eaten. Had entertained just two visitors: his brother Jack and his dad. There wasn't much to say.

Van had a final question: "Was there anyone else in the accident from around here?"

"No," Mary Jane answered. "They were from Cache Valley."

THAT NIGHT, SHE AND Ed talked about the questions Rindlisbacher was asking—suggesting Reggie somehow was at fault—and they were worried sick. Ed didn't express it like his wife. But there was a story he'd heard about, maybe read about in the paper, that was running around his head—making him nuts.

It was this story about a kid in Idaho—two hours away, and in so many ways Utah's cultural and political sibling. Not too many months before, Ed remembered that there had been this young man, eighteen years old or something like that, who had been arrested. He hadn't committed a big crime, just some silly misdemeanor. But the cops had picked him up. Then they called the kid's dad to come get him. The dad said: "Let him spend the night in jail; it'll be a good learning experience for him."

"In jail that night, some guys beat the kid to death," says Ed. As he recalls his thinking at the time, his eyes mist. Then a tear rolls down his cheek and he clenches his teeth. "All I could think was: 'I cannot let Reggie spend a single night in jail.'

"'I will sell the house. I will do whatever it takes. But I will not let him spend a night in jail.'"

JACKIE HAD BEEN NUMB all day at work, knowing the worst was to come. She had to tell the girls what had happened to their dad. After receiving news of the accident, she'd stayed at work until three in the afternoon. Then she went to school to pick up her oldest daughter, Stephanie, a real daddy's girl. Jackie's colleague Roy drove her in her own car over to Thomas Edison Charter School.

Jackie gave the news to the principal and to Stephanie's teacher. Then she walked with Stephanie to her Saturn—just like Jim's, a practical car, navy blue. Stephanie had on her school uniform, the khaki slacks and polo shirt, her blond hair in a protective braid because it was gymnastics day.

Stephanie saw Roy, her mommy's work friend, and asked: "Are we going to gymnastics?"

"Sweetie, there's something I want to talk to you about."

Jackie sat down in the front seat and pulled her seven-year-old onto her lap. She realized Stephanie didn't have too much understanding about death, or, at least, she didn't have much experience with it, except for when they'd put Sandy, their chocolate point Siamese cat, to sleep two years earlier.

"Daddy was in a car accident."

Stephanie looked at her with those deep blue eyes. She never thought she'd have a blond-haired, blue-eyed girl.

"He didn't make it."

Stephanie started to cry and now Jackie did, too. Stephanie curled up in her mother's arms. "I was holding her, and I was hiding her." From the world, everything.

"He's gone to heaven."

They sat there like that for a while, Stephanie not speaking, not even asking questions. She didn't ask any questions. Jackie and Roy got her into the backseat. They drove to the day care center to get Cassidy, who was just a few months shy of four years old. On the way, Stephanie finally offered up a question: "So are we going to do what we'd planned this weekend?"

It surprised Roy, the sophisticated implication of the question: Will life go on as usual? A few hours later, people had begun to fill the Furfaro home. The phone rang. Jackie's mom picked up. It was a reporter from the local paper. Jackie heard her mom say, "How the hell do you *think* we feel?" and hang up.

Jackie tried not to make a public show of her grief, trying to eat, not eating, shaking, trying not to shake, and thinking: *How am I going to do this alone?*

It was already dark when Jackie carried Cassidy to the living room just inside the front door. A big TV stood in the corner. And there was a massage chair, from Sharper Image. It cost $2,100, but it was worth it to help with Jim's periodic migraines.

Jackie and Cassidy curled up on the dark tan fabric couch. The little one was a mamma's girl, who liked to lie on Jackie's chest. By then, Cassidy had asked when Daddy was coming home, and Jackie was trying to find the right moment.

On the couch, she tried to calmly explain that Daddy had had an accident and had gone to heaven. "He isn't coming home."

She wasn't sure how much Cassidy understood. And, if Jackie was honest with herself, she wasn't sure whether she even believed it. "A small part of me thought: 'Maybe it's a mistake, and he'll come home anyway.'"

ABOUT THAT TIME, OVER in Logan, a woman named Terryl Warner pulled up in a blue minivan outside a gymnastics center called Air-Bound. The door pulled open and Terryl's daughter, a sixth grader, chilly from waiting outside for a few minutes, climbed inside.

"You won't believe what happened," the girl, Jayme, started. "Cecily told us that Stephanie's dad was killed in a car accident."

Terryl paused and tried to catch up to the conversation. Cecily was the gymnastics coach.

"Which one is Stephanie?" she asked.

Jamie reminded Terryl that Stephanie was on her gymnastics team, the daughter of Jackie and Jim Furfaro. Terryl began to form a picture. She'd had modest interaction with the family but liked them. She remembered, three months earlier, the night before a statewide meet at Air-Bound, when Jim Furfaro and her husband, Alan, had to set up the fiberglass springs under the mat. They'd cut their hands on the fiberglass and had to go out and get gloves to finish the job.

Meantime, Terryl had been trying to make some posters for the gymnastics meet. The posters had flames on them, but Terryl's drawings were so bad that they'd never be used. They were so awful, in fact, that Jackie had laughed in a friendly way at them, something Terryl appreciated; she tried not to take herself too seriously.

Terryl was shocked, of course, by news of the wreck. But she was used to hearing about tragedy. In her daily life, Terryl came across a lot of horrible situations. She was the victim's advocate in Cache County, which encompassed Logan. It was a particularly low crime area, but it still had its share of particularly gruesome crimes, like rapes and child abuse. In Terryl, the criminals had a particularly ardent and zealous foe. She had a reputation for relentlessly pushing for justice, even among the prosecutors.

Terryl connected to people who suffered tragedy, in a very personal way. She spoke from experience. She'd been on her own improbable journey, a victim from a very early age, and she'd learned to fight back—for herself, for others.

When her daughter told her about the car crash, Terryl didn't think about it in professional terms. After all, it sounded like it was just an accident. Besides, Terryl assumed it must've happened in nearby Box Elder County, which was home to both Tremonton and ATK Systems. Not her jurisdiction.

Mainly, Terryl thought: *Jackie's going to be a single mom. What a terrible tragedy.*

CHAPTER 5

TERRYL

IN JUNE 1980, IN a room with bedsheets tacked up as curtains, a slight high schooler with brown hair was fast asleep when she recalls she was awakened by yelling: "Get out of bed! Get in here! I want you to see this."

It was the summer after Terryl Danielson's ninth-grade year in Downey, California, a town near Compton, not far from Los Angeles. A tough neighborhood, but no tougher than what happened inside the walls of the three-bedroom house with overgrown ivy in the front yard that Terryl, her mom, Kathie, and older brother, Michael, had given up trying to groom.

"Get in here!" Terryl remembers her father yelling as she fought to get her bearings. She could see the .357 Magnum in his hand.

"I'm going to blow your mother's fucking head off."

THINGS HAD BEEN LEADING to this moment for a long time, a slow-motion wreck starting in August 1962 when Kathie, who was chunky and adorable with short blond hair and had just graduated from high school, gave birth to Michael. He was a big kid but passive. Less than a year later,

Kathie delivered again, this time with twins, Terryl and her sister, Kerryl.

Several months after she was born, Kerryl died. They chalked it up to pneumonia.

Dad was Byron Lloyd Danielson. He built driveshafts at a local garage. He was big, six-foot-two and 220 pounds. He was clean-shaven and kept his brown hair cut short. He was good-looking. He could be really sweet. There was this one time that he and Kathie and a bunch of their friends went to an Elvis concert in Long Beach. There weren't enough tickets for everyone, but he managed to score a really great one for one of Kathie's best friends, a better seat than anyone else's in the group. Then things began to change for Danny—as the family called him—when he hurt his back at work. He started taking painkillers. And he drank. His personality changed. When he was drinking or popping pills, he'd go from "Dr. Jekyll to Mr. Hyde, from Joe-Mr.-Nice-Guy to a real a-hole. You couldn't believe it was the same guy," recollects Nanci Smith, a close family friend. She remembers Danny always having guns, keeping one stored beneath the driver's seat of his car.

TERRYL'S MEMORIES OF TERROR start early. In the year before she started kindergarten, she was asleep in the family's previous house in nearby Lynwood, in a room with yellow walls that she shared with Michael, when she was yanked out of bed in the middle of the night. "My mom's face was bloody." They ran across the street to a neighbor's house. "It was my first memory of violence."

The family called Byron "Danny" after his last name, Danielson. Sometimes Dad, but often Danny. They stayed as far away from him as they could, as their tiny house would permit.

There wasn't much of anywhere else to go, given the gangs and toughs who roamed the streets. Terryl retreated into books. Anything she could get her hands on. She loved the Bobbsey Twins.

Kathie was a Mormon, and she would sometimes take the kids to a nearby LDS church. But Danny, not a Mormon, wasn't a fan of the activity. Terryl remembers that Danny, to make sure church wasn't an option, would sometimes take the distributor cap off of Kathie's long, chocolate brown Cadillac, disabling the aging sedan.

Kathie tried to keep up outward appearances at the Lynwood house. She would borrow a lawn mower from a neighbor, and she and Terryl and Michael would groom the front yard.

One afternoon when Terryl was in the fourth grade, she remembers finding a cup of orange juice in the kitchen. She took a sip and discovered it wasn't orange juice, or, rather, not just orange juice. It was what her dad called a screwdriver. Orange juice mixed with the vodka that made him crazy and violent. She dumped out his bottle of vodka. Come what may for Terryl, the smallest girl in her class.

"He came in screaming, and he grabbed me by the arm. He dragged me to my room, and he spanked me with a belt. He was in a rage," she recalls. It hurt, a lot. It didn't stop her.

"Every time I could throw out his vodka, I did."

She learned very early on to disassociate herself, to lock out the emotions. She says she didn't cry when he beat her. Her brother Michael cried, but not Terryl.

Each morning and night, Terryl knelt beside her bed. Every night, the same prayers: *Why couldn't I have been born into a family where my parents loved each other and were happy for their kids?*

She begged God to help her mother leave her dad.

Terryl recalls that Kathie had a different plan. She decided there was a way to make things better: relocate to a new place. So when Terryl was nearing the end of elementary school they moved to a three-bedroom apartment in Whittier, another tough city, and they lived in the rough part of it. And now the liquor store wasn't a few blocks away, it was directly across the street.

Terryl had her own room and so did Michael. Now they had a little brother, Mitchell. The kids didn't leave the apartment because they feared the gangs. Terryl didn't leave her room because she feared her dad. She escaped further into books. Nancy Drew, the Hardy Boys.

There was a mild mystery in the house. Terryl says she was told not to answer the phone. She didn't think too much about it. It was just a rule. *Don't answer the phone, Terryl,* and she didn't. Better to follow the rules, take them on faith, not upset the delicate balance more than it would get upset on its own.

During that period, she kept a diary. She had a looping cursive that stood straight up. The entries would juxtapose the mundane thrill and confusion of being a young girl with the terror of living with Danny. On November 16, 1977, a few years before the gun incident, she wrote: "I talked to Greg Hertzberg yesterday, it was so good to talk to him! I really love him! I failed a math test today but hope to get a good grade on the test tomorrow." And a few sentences later: "Mitchell can walk at nine months now and tonight we took movie films of him walking. Last night, I only got four hours of sleep. Dad got drunk and started playing Michael's saxophone and would not stop. It was dreadful, he woke everyone and said that the reason he was playing it in the middle of the night was because that was the only thing or person or whatever that understood. Mom said she was glad something did. Tonight, it is hidden under my bed."

The months wore on. More fights. In January, she scribbled, "I am writing this in the dark because I'm afraid of what Dad will do if he sees that the light is on."

MICHAEL STARTED TO GET into drugs, Terryl explains. Pot, when he could get it. When he couldn't, Terryl watched him sniff gasoline. After taking a big whiff of the gas, he'd pass out. The cycle of addiction was beginning to take hold. And Terryl felt the other ignominies in her life, such as sometimes showing up at school without a lunch.

"Do you have anything left over?" Terryl recalls asking classmates in the cafeteria. "Can I have some of your chips?"

Terryl and Michael sometimes would get silver dollars from their grandparents. When they were flush with a silver coin or two, they'd go to the local convenience store and buy food for breakfast or dinner, often splitting a pack of Hostess donuts.

One evening, when they were living in the new apartment, Danny came home loaded, in a rage. He threw Kathie and Michael out of the house. Then he came for Terryl.

"You drunk son of a bitch. You are worthless!" she remembers screaming. He chased after her and grabbed her. She kicked and punched. "You are not getting me out of this house without a fight!"

He got the better of her, bloodied her, put her out with the others. Terryl wrote in her diary: "This man cannot be my dad. I can't have come from someone this evil."

She was furious at her brother, too.

"Go fight back!" she says she implored Michael. He was a big kid. Not Danny's size, but big for his age. "When he comes after you, fight back."

Her impression was that Michael and Kathie were peacemakers who wanted to stay under the radar. And that her mom felt she could change the dynamics by again changing the circumstances and the setting. She had a new plan: move to Downey, a new community. A new start.

It was the summer after Terryl's ninth-grade year. Her room with the sheets tacked up for curtains also had a built-in desk, a chest of drawers, and a record player given to her by her grandparents. She had a Steve Miller album, but she wasn't much into music.

Now she was an even more voracious reader. In fact, she'd gotten herself into trouble for the first time. The reason: She'd checked out too many books. Her favorites were *The Diddakoi*, the story of an orphan girl who faces persecution, and *The Secret Garden*, where a girl and a boy, who is sick, escape to a beautiful garden. She loved mysteries. She hated teen romances. She'd started to get into Stephen King. From the Downey library, she would take out armfuls of books, fifteen or even twenty at a time. She didn't always return them on time. She got a fine: $300. She couldn't pay it.

But for as much of an emotional escape that reading provided, it couldn't protect her from the physical threats. It was not a sufficient one on that summer night when Danny burst into her room with a gun.

"GET IN HERE. I want you to see this!"

She was wearing a nightgown. She and Michael followed Danny's orders. Michael had come from his room and started to cry. Mitchell, their baby brother, was back in Michael's room, sleeping through it all, as he'd learned to do. Terryl and Michael followed Danny into their parents' small bedroom. It had hot-pink walls. Kathie lay on the bed, her face streaked with mascara. Danny walked over to her. Terryl and Michael stood in the doorway. Now Terryl was crying, too.

"Please, please put the gun down."

Terryl remembers her dad standing over her mom and pointing the gun down at her. Pushing it toward her mouth.

"Please," Terryl says she and Michael begged. Kathie sobbed.

Danny took a step back. He lowered the gun.

CHAPTER 6

THE TOLL

AFTER A HARROWING DAY, it was Rindlisbacher's habit to talk about it, not bury it deeper. Things were so-so with his wife, Judy. He'd not been able to give the marriage quite the attention he'd have liked, having spent so much time in the military, traveling, and otherwise in and out of the home. Two of their kids were grown and out of the house. But, that night, he sat down with his seventeen-year-old daughter, Allison.

"Please slow down. Please wear your seat belt. Please don't be stupid when you're driving a car."

That night, his mind was racing with the events of the day. After he'd taken Reggie to the hospital, he'd escorted Kaiserman and the farrier's wife back to the scene. Then he'd driven the Crown Vic ten minutes to the Cache County Sheriff's Office, a new building that housed the local law enforcement, and was next to the jail. In the third-floor offices, Rindlisbacher typed in the witness statements. He loaded the photos from his personal camera to revisit the scene. In the photos, Jim's head lies back, his eyes closed, a crisscrossing of red blood across his head more suggestive of a bar fight than a fatal wreck. He's got short-cropped brown hair and a goatee. He looks to Rindlisbacher to be at peace.

Not so much Keith. He'd taken the brunt of the impact from Kaiserman's load. In the photo, Rindlisbacher could see Keith slumped forward, a hint of male pattern balding on top of his head. And there, in the back of the Saturn, a mass of pink and gooey stuff that had been sprayed on impact. Keith's brain.

As Rindlisbacher retired for the night, he was juggling things he couldn't get out of his head: Kaiserman's statement that Reggie had crossed the yellow divider several times prior to the crash; Reggie's texting during the ride to the hospital; Reggie's inability to offer any other explanation for crossing the yellow line.

"If he'd said he was tired, I might have left it at that," Rindlisbacher says, looking back. "He could've lied to me and I'd have had nothing to refute it."

"I WANT TO SEE him."

"Ms. O'Dell, why don't we take care of this paperwork first?" the mortician said. It was the day after the wreck. Leila had slept little, despite taking a sleeping pill around eleven p.m. During the day, she'd been inconsolable, barely able to speak to the family members who came by, sobbing with her daughter, Megan.

Megan left in the evening. She had previously signed up for doing roadside security for the Top of Utah Marathon, which was being held the next morning. But she had to be at her post that night, spending much of it guarding runners' possessions.

Still, when Leila got to Allen-Hall Mortuary the next morning, she was hoping to see Megan. But the young lady had slept in. Even absent the marathon, it was Megan's habit to stay up late, often playing video games, and then sleep late.

Leila was determined to see Keith's body. But the staff was clearly trying to stall, and even distract her. She complied with their request and gave them particulars, like Keith's Social Security number and a bunch of dates and names she rattled off. She selected a casket, something in a medium-colored oak that she thought Keith, a fan of natural wood, would like. She picked sunflowers, which she thought was like Keith, not roses or carnations.

They asked when she'd like to have the funeral. It was Saturday. They decided: Wednesday. And they decided to do it the same day as Jim's funeral—one following the next—because they had so many friends and colleagues in common.

Leila asked again to see the body. Someone made it plain to her why she couldn't see it. Keith, her true love, was broken beyond anything imaginable.

A man handed Leila a Ziploc bag. It contained a broken cell phone and a GPS, his watch, his car keys on a chain with a small compass on it, and the brown leather wallet he had made in his high school shop class and carried ever since. In another bag, she received his black lace-up shoes and his dark socks.

"Where are his glasses?"

Gone. This was all of Keith that remained.

She couldn't bring herself to eat. She didn't pray. She had become something akin to a zombie. More visitors came to the house.

Megan showed up. Her life was already out-of-balance and now her dad was dead. When she was a little girl, she and Keith would go geocaching together. He'd wake her in the middle of the night and they'd use the big Deep Space Hunter telescope to look at a comet or other celestial event. They would throw the softball around in the yard. They played video games, the first being an early *Star Wars* game that they could play against each other after Keith used his wizardry to connect two computers together—an early, jerry-rigged, multiplayer game.

Megan looked up to him. "I wanted to be a rocket scientist, like my dad. That's what I wanted to be, if I wasn't going to make the Olympics."

But things started to sour for Megan, and in her relationship with her parents, in high school. Maybe it was because she switched to a bigger school or maybe because she got her first boyfriend. Her grades dipped into the regular B's, then got worse. She swam less. Then, partway through high school, she said she was raped by a boy who lived near the big church. The boy had plied her with beer, she recalls. Her recollections were fuzzy; they came in and out. In the end, nothing came of the allegations; she felt her parents weren't supportive.

Her grades worsened. She stopped swimming, thanks in part to an injured shoulder. On her high school graduation night, not long before the accident on Valley View Drive, her boyfriend got down on one knee at a campsite and proposed. Megan went home to show her parents the ring; she said her mom didn't approve. On the other hand, because of their souring relationship, Megan had asked her mother not to come to her high school graduation. As to her engagement, she said, her father seemed mostly supportive. "I don't think either of my parents liked him," she said of her fiancé. "My dad kind of went with it."

Megan thought of her father as a tether, and he was gone.

She needed something to wear to the funeral. She was low on funds. Leila looked inside Keith's wallet; as she suspected it would, it kept $100 in $20 bills—what Keith usually carried. She gave it to Megan to go with her aunt to the mall to buy something black.

Among the visitors to the house was Tom Higgs, Keith's boss—*putative* boss, really, given that Keith was the guy at the office answering questions. When Tom got to the O'Dells', he saw Leila in the living room, shaking.

"She couldn't focus," Tom recalls. It was like trying to talk to someone who didn't have a reason to take the next breath.

"CAN WE SEE HIM?" asked Jim Furfaro's mom. She meant: Jim's body.

The mortician nodded grimly.

Jim's mom stood behind Jackie at Allen-Hall Mortuary. The Furfaro family made their visit shortly after the O'Dell clan.

Jackie already had her game face on, an almost impassive look, the grief held tightly inside. After falling apart, turning hysterical, when the officers first told her in the classroom at Utah State, she recovered her outward composure.

"In the end, I was embarrassed that I'd fallen apart in front of anyone who knew me. That isn't my thing."

After she told Cassidy, she put Stephanie to bed. The older girl, upstairs in her room, attempted to be brave herself, tried to sleep. She could hear the relatives downstairs, talking about it. She had put her hands over her ears and held them tightly, trying to make it all go away.

A bit later, Jackie climbed into bed and pulled Jim's heavy terry-cloth robe over herself. Eventually, unable to sleep, she took NyQuil. And then there was a pitter-patter of feet, and Stephanie crawled into bed and they sort of slept the rest of the night under the robe that smelled like Daddy.

In the morning, Jackie disengaged from the family that had gathered to pay respects and take her to the mortuary. She went into the bathroom, got in the shower, and sobbed. Heaved. Leaned against the wall, let the grief out. She reappeared thirty minutes later, drained, back in a modicum of control.

Then, at the mortuary, they put her through the same paces as they had Leila, just an hour earlier. Jackie filled out paperwork and picked out a casket.

"Can we see him?"

Jackie wasn't particularly religious. Jim was a confirmed Catholic. But he and Jackie weren't into anything organized. Jackie had a general belief in God, tempered by a scientist's view of the world. As often as not, though, she relied on herself. She was the rock, a role she'd always taken to and sort of relished.

"I always felt like I was the shoulder."

Jackie walked into the viewing room. Jim lay on a gurney, covered with a comforter. Not a sheet, which struck Jackie, but a comforter, light-colored with a flower print. They pulled back the cover.

Dried blood trickled from Jim's ear. His right eye was missing. She touched his chest and head and hair. He was cold. She thought: *He looks pretty good, given what they told me about the accident.*

She whispered: "Good-bye."

CHAPTER 7

THE NEUROSCIENTISTS

I'M A TREE FREAK."

A twenty-foot-high triangle palm stretches up from its pot in the middle of the main floor of Dr. Gazzaley's studio loft. The palm divides the kitchen from the living room, which has a sleek black couch and a small Japanese maple in the corner.

The couch faces a gas fireplace, a light blue flame burning. Over it hangs a glossy painting called *Three Oaks* with three trees floating in the air against a yellow backdrop. The puffy trees look like brains.

Alternative rock with an electriconic beat pulses through the loft.

"The band is called Metric," says Dr. Gazzaley. It's a new track, which is one of his rules when it comes to music: He won't put a song that's been out for more than a year on the playlist. "I'm very into new bands."

Dr. Gazzaley pushes a button on the fireplace. A screen descends until it covers the top half of the fireplace, including the three oaks. It displays high-definition images shown from the projector hanging from the ceiling across the room. It's also connected to the PlayStation 3 so that he and his girlfriend, Jo Fung, can play video games, their current favorite being Assassin's Creed.

Checkerboard windows stretch from the floor to nearly the loft's roof. They look to downtown San Francisco on a clear, chilly night. The street below the loft hums with activity, twenty-somethings giddy on martinis. This is the Mission District, a gentrified hotbed of sushi, tapas, hip Mexican restaurants, and upscale furniture stores.

Inside the "Gazzloft," the energy is starting to heat up. Tonight marks the party's fourth anniversary, and the host certainly wouldn't miss this Friday, even though he's already logged a huge day.

It began in Florida, where yesterday he spoke to a group—"the college of neuropsychopharmacology, something like that," he says. He could be excused for being tired. He got up at three a.m. California time. As soon as he flew home, he met an NBC crew, which asked him to scan the brain of a multitasker for a segment for *Dateline* and the *Today* show.

"I'm always in a hurry," Dr. Gazzaley says. And he doesn't do anything halfway. "I don't like to take on anything I can't blow away."

He says his drive comes from a passion for science that had him, as a kid, taking the bus two hours a day to and from a working-class neighborhood in Queens, New York—the son of blue-collar parents (mom was a bookkeeper, dad worked for the transit authority)—to the world-class Bronx High School of Science. Despite excelling there, and getting high SAT scores, he didn't get his first college choices. It intensified his drive, made him want to prove himself, but, he says, it also freed him to approach problems in his own way. "The system failed me, so I decided I didn't need the system."

He feels a buzz from his phone. He extracts it from his black jeans and reads. "Oh, look at this!" He holds up the phone and shows the text: *I'm coming. Look out!*

It's from Philip Rosedale. "He created Second Life," Dr. Gazzaley says. It's one of the biggest virtual worlds ever created.

Dr. Gazzaley walks to the kitchen counter to show the text to a handful of people who have come early to eat prosciutto and soy chicken from Whole Foods. "It's going to get crazy," Dr. Gazzaley says. In addition to Rosedale, he's expecting a bunch of hipsters, rockers, and entrepreneurs: the guitarist from Counting Crows and the lead musician from Thievery Corporation; the guy who created Digg (an Internet service that hundreds of millions of people use to rank whether or not they "like" something on the Internet); owners of some of the swanky nearby restaurants; and a gorgeous woman from the Caribbean named Rio, who professes to "be in love with Dr. Gazzaley's brain."

BEFORE THE PLACE EXPLODES, Dr. Gazzaley says there is time to explain the "cocktail party problem." It is among the most fundamental precepts in attention science. As he talks, he makes a show of shifting his eyes to one of the handful of pre-party attendees, a pregnant woman named Kat who oversees HR for Burning Man.

He explains that he's illustrating the cocktail party effect.

"I can look at Kat and still be listening to whoever I'm talking to," Dr. Gazzaley says, then shifts his eyes back to me. "Or I can look at whoever I'm talking to but shift my attention so that I'm listening to Kat.

"It's an exercise in selective attention."

What he means is that people have a powerful, even extraordinary, ability to direct their attention to control it. At the same time, he says, the cocktail party effect shows the limitations of attention; after all, you can't pay attention to two conversations at once. In fact, it's so limited that if you're really listening to the person in front of you, there are generally only two things you can pick up in a different conversation: the gender of the person speaking or, in some cases, the sound of your name.

"There's an illusion that you have a whole field of attention," Dr. Gazzaley says. The reality is that you can focus your attention on one very specific thing, not everything in your field of vision. Dr. Gazzaley says the meaning of this principle is actually open to interpretation. On one hand, it shows you can control and focus your attention. But seen another way, the exercise shows how hard it is to spread your attention. It is more like a laser than an overhead light.

Is that good or bad? "It depends on whether you're an optimist or a pessimist," Dr. Gazzaley says.

In short, attention is extremely powerful and extremely limited. That idea, as embodied by the cocktail party effect, is fundamental to modern attention science, and it seems so obvious. But it was quite revelatory in the middle of the nineteenth century. Up to that point, there was a general belief that the brain was "infinite" in its power to take in and process the world.

IF YOU TOUCH YOUR foot, you feel it. Immediately. Right?

Until the middle of the nineteenth century that thinking symbolized the prevailing scientific thought about the human neural network. In a nutshell: Scientists believed that a human's reaction time was immeasurably fast—infinite, as some scientists put it. A person felt a prick and instantly recoiled, or saw a snake and leapt, or heard the sound of thundering hooves and stepped back off the street.

What could prove otherwise? Try a "galvanometer." It was an early device used to detect and measure the magnitude of an electric current. Such gadgets were also employed to measure speed; how fast was a current traveling? The devices took various forms, but basically they used twisted metal coils to detect an electrical signal. When a galvanometer detected a signal, it created a magnetic field. The magnetic force, in turn, would cause a little pointer to move, allowing a researcher to measure when a current first appeared, and when it dissipated. Early researchers used them in the field of ballistics to, say, measure the pace and trajectory of a bullet.

Hermann von Helmholtz used one to measure the reaction time of a frog.

Helmholtz was an associate professor at the Prussian University of Königsberg and one of the leading scientists of his day. He used an electric current to stimulate an amphibian's calf, then employed the galvanometer to measure how long it took the electrical current to run along the sciatic nerve. What he found changed the way scientists viewed the human brain. He discovered that what had looked instantaneous from the outside, was, in fact, anything but, and not even close to light speed. It took time for the current to travel to the brain.

He tried it on a human being and discovered the same thing. Roughly, he estimated that "neural conduction time" was around one hundred meters per second. It took twenty milliseconds for information to get from brain to calf. (By comparison, light travels at about 300,000 kilometers per second.)

Helmholtz's use of the galvanometer symbolized not just how scientists were measuring human capacity, but why. The researchers realized that the machines people were building—those mechanized tools of productivity, war, and science—were so powerful that they were moving faster than people could keep up. This was a defining moment. Scientists were putting new technology to work to understand the capacities and limits of the human brain at the very moment that technology was putting those limits into sharp relief.

"It was not obvious that people were slow to respond until the 1850s. The increasing use of machines began to make this obvious," says Michael I. Posner, a professor emeritus of neuroscience at the University of Oregon, who is one of the modern pioneers of attention science and a student of its history. The relatively slow reaction time of humans "wasn't clear until the pressure on people started to rise."

Through the nineteenth century, technology quickly evolved. In 1837, across the ocean, Samuel Morse unveiled his first telegraph device, and five years later "Morse convinced Congress to provide $30,000 in support of his plan to 'wire' the United States," according to a history compiled by the Massachusetts Institute of Technology. It notes that in May of 1844, Morse gave a public demonstration, sending a message from Washington to a train depot in Baltimore. The message read: *What hath God wrought?*

There were twenty-three thousand miles of telegraph wire in operation by 1854, MIT reported (Western Union was founded in 1851). In

1866, a cable link connected the United States and Europe. Thanks to machines, data was moving much faster than humans could ever transport it.

In Europe, another innovator, Charles Babbage, born the son of a banker in 1791, was developing an early calculating machine at the time that Morse was refining and evangelizing the telegraph. Babbage designed schematics for a programmable computer, something that could process mathematical equations in place of the human mind. He was celebrated at the time for these marvelous concepts—and then much later as one of the computer's great pioneers.

As the technology evolved, it had almost competing roles. It was simultaneously leading to tools that would make humans so much more powerful, and at the same time, it was making humans seem less powerful, certainly slower, than they'd ever been.

A boy from the same hometown as modern-art master Vincent van Gogh was about to show just how slow.

TILBURG IS A MANUFACTURING town in North Brabant in the Netherlands, known today as the place Van Gogh did his childhood schooling. But in 1818, more than two decades earlier, Tilburg saw the birth of Francis Cornelius Donders. F. C. Donders learned early in his life to pay attention; he had eight older sisters. He would later study attention with unprecedented precision.

Like Helmholtz, he developed into a scientist with many interests, though he was known early in his career for ophthalmology and for establishing the first eye hospital in the Netherlands in 1858. In the middle of the 1860s, Donders turned his attention to human reaction time. He came up with an experiment that involved sitting two people in front of a "phonautograph." It was an early recording device, patented in 1857 by a Frenchman, that captured sound and then, through a mechanical arm, transcribed it onto a piece of paper. In Donders's experiment, one participant uttered a syllable, then the other repeated it as quickly as possible.

Using the phonautograph, Donders was beginning to measure what he called "the timing of the mind" and "mental action."

He set about trying to determine what circumstances led to shorter and longer reaction times. To do so, he employed three simple tests. In one, a person sat in front of a panel of lightbulbs and pressed a button when a light went on; in a second, the person faced a slightly more complex task of hitting a specific button corresponding to a specific light; in the third, the person was required to hit a button when one of the lights went on, though this time if the person failed to do so in a timely manner, the other light would go on.

There is a seemingly obvious quality to his conclusions: Simple tasks take shorter time and that time grows with the complexity and type of process involved. For instance, the reaction time grows if the person must make a choice, or engage a particular motor skill, like pressing a specific button.

Donders expanded the experiments. He introduced a range of different demands and stimuli, asking lab participants to respond with the left hand, or the right, and to discern among and between colors, words, sounds. He discovered that more complex tasks not only took more time, they introduced a new wrinkle: more error.

How quickly could the brain work, and how much information could it take in and at what speed? Donders built instruments to try to measure what he called "the time required for simple mental processes."

"He was measuring," Dr. Posner explains, "the time it takes to have a thought.

"As you go into the history of the study of attention, the really epic discovery was the speed of mental processing."

Donders's writing at the time shows an ebullient quality, such as when he describes the challenges of quantifying mental processing. "But will all qualitative treatment of mental processes be out of the question then? By no means!"

There was also an ominous note to his findings, published in 1868. He wrote: "Distraction during the appearance of the stimulus is always punished with the prolongation of the process," notes a biography published by the Donders Institute for Brain, Cognition, and Behaviour in the Netherlands.

Meantime, computer technology, such as it was, was getting more sophisticated. The cutting-edge computer of the period would hardly have been seen as distracting, unless by virtue of its sheer size people couldn't take their eyes off of it. Take, for instance, the Hollerith tabulator. It looked sort of like an upright piano, and used punch cards. Thanks to use of the device in 1890, the calculation of census data at the time fell to three years, down from seven, according to the Computer History Museum. At the time, a publication called the *Electrical Engineer* remarked that the tabulator bested even divine speeds. "This apparatus works unerringly as the mills of the gods, but beats them hollow as to speed," the *Electrical Engineer* read, according to a history published by Columbia University.

The Hollerith tabulator was the work of Herman Hollerith, who incorporated the Tabulating Machine Company at the end of the nineteenth century. Later, in 1911, it became the Computing-Tabulating-Recording Company, which in turn was renamed in 1924 as International Business Machines, or IBM. Modern computing was in its infancy, still used only by corporations and the military, or other huge government operations. But in those circles, its power was becoming apparent.

On a mostly separate track, there were also major advances in communications technology. Alexander Graham Bell, working on development of a talking telegraph, got essential patents toward his efforts in 1876 and 1877, the year he and partners started the Bell Telephone Company. By 1904, there were 3.3 million telephones in the United States, according to a corporate history published by AT&T. Then, by 1927, transatlantic telephone calls were available using two-way radio technology, with a call costing $75 for three minutes.

As the twentieth century progressed, developments in computing and interpersonal communications technology, which had been evolving somewhat independently, began to converge. It is a powerful union whose significance cannot be understated. The marriage of computers and communications brought so much utility. It also set the stage for a formidable, arguably unprecedented, challenge to the human brain, one that was often unseen and widely underestimated. Put another way, technology was evolving by the day, but the human brain was more or less staying put.

BY TEN P.M., THE Gazzloft is hopping. The host, enjoying pats on the back and gifts of booze from newcomers, slides through the crowd and sidles up to his girlfriend, who in turn is talking to a man with a three-day growth, a friendly grin, and a gray fedora. Dan Vickrey is the guitarist for Counting Crows and cowriter of some of the band's big hits. He's just back from tour, his face ruddy from the effects of days on the road. These days, he says, bands make their money from touring, not from putting out music, even hits. That's because the Internet has shattered traditional business models and fractured the attention of the audience. So many entertainment options, so little time.

While on this latest tour, he explains, he had a cool experience related to technology. In the front row at one of the concerts sat a cherubic man with a white beard. Dan didn't recognize Steve Wozniak, the cofounder of Apple Computer. But he was wowed when, later, after the concert, Woz came backstage and asked Dan if he wanted to have dinner sometime, just to hobnob and hang out. "How cool is that!" Dan says.

"Look!" Dr. Gazzaley exclaims.

His girlfriend and Dan follow the neuroscientist's gaze to a group of three people standing nearby in the middle of the packed room. One of the three clearly commands the group's focus. He's tall with hair that is both short and unkempt. On his nose sits a pair of odd-looking glasses. They look sort of like the slimmed-down opaque wraparound shades worn by cyclists.

"Google glasses. One of ten pairs in the world!"

The glasses are not used to help show a person the outside world but, rather, a flow of information. When a wearer glances to a spot in the corner of a lens, he or she can scan a data feed to, say, check incoming email. In this case, they are mere prototype, a grand curiosity being worn by another local celeb, Philip Rosedale, the founder of Second Life.

After the Google Glass sighting, Gazzaley, his girlfriend, and Dan the guitarist, each pull out their cell phones. Dr. Gazzaley looks up to see the others lost in their own space. "That's funny. That's a moment," he observes. He surmises: Each of us was probably going to tweet about it or post the moment on Facebook.

Dr. Gazzaley says various media outlets have been calling him to ask about the new Google glasses and how they might impact attention.

"They're distracting! I keep telling them." But, he says, "They don't want to hear it. They have an agenda, and they're going to print whatever they want."

THIS IS A SIGNIFICANT moment. Here, amid the digerati, the elite of the Valley, the definition of attention seems to broaden. It is not merely about the cocktail party effect, which demonstrates the power and limitations of our ability to focus. In this setting, amid a cacophony of face-to-face and virtual communications, another key piece falls into place. It has to do with our desires and efforts to capture other people's attention.

After all, all the tweets and Facebook status updates, the emails, YouTube videos, and texts are not creating themselves. We are creating them. They are enabled by technology, sure. But they are driven by the humans pressing the buttons, asking for a tiny piece of the fractured spotlight. In modern life, as at this cocktail party, the noise is not incidental, not a laboratory exercise. It is everywhere, and it is created by someone, or many, each with their needs.

In fact, the ability for individuals to create and broadcast media explains a key difference between the nature and amount of information people consume today than in past eras. In 2008, people consumed three times the amount of information they did in 1960, according to researchers at the University of California at San Diego. And now, the researchers say, one-third of the information we consume is interactive (as opposed to passive media like television or radio).

In the spring of 2013, a few months after this First Friday party took place, Twitter announced that users were sending 400 million tweets per day. That's up from 340 million per day a year earlier, a 17.6 percent increase. That's a mere drop in the bucket, compared to the number of texts sent—*six billion* in the United States each day. Email adds another 144 billion globally on a daily basis. And don't neglect Facebook, which reported in the summer of 2012 that users were daily posting 2.5 billion pieces of content (status updates, photos, videos, comments, etc.).

The cascade also spills over into the work world. A typical office worker was getting interrupted by various media stimulation every three minutes in 2004, according to research by Gloria Mark, a professor at the University of California at Irvine. That was before the spread of instant messaging and Facebook. By 2013, the interruptions were every two minutes; such intrusions came either from a person responding to a new stimulus—like an incoming email—or from an internal urge to change tasks, say, to write a new email.

Dr. Gazzaley sees another side effect in that there were so many medical journals popping up online that it was getting hard to keep up with all the new developments and to figure out which were valid and which were not.

"More journals are coming into existence. It's hard to keep pace with the articles hitting journals. It's almost impossible, there's so much of it," he says. He's a full participant, publishing articles, speaking, doing media appearances. Broadly, the challenge of keeping up is "a problem of oversaturation." At the Gazzloft, Dr. Gazzaley attracts the attention of the old-fashioned variety: A beautiful woman with long black hair, dressed in a tight skirt, gives him an adoring eye. Before he peels away to hobnob, he says he has a thought about another way to illustrate the way attention works, and how easily it can be fractured in the modern world. He wants to show me the power of distraction.

Through Mickey Hart, he explains, he's met a man named Patrick Martin. Patrick is a magician. "He's done magic for heads of state," in addition to Muhammad Ali and Princess Di.

Dr. Gazzaley says the magician is a master of manipulating attention. In a room in which he is exposed to an audience all around him—360 degrees—he manages to steer the attention of everyone exactly where he wants it, and away from the thing they think they are trying to focus on: the sleight of hand.

Of late, the magician has Dr. Gazzaley thinking about distraction. It's not exactly the opposite of attention. But it is an antagonist to attention. As such, Dr. Gazzaley thinks the concept of distraction provides a powerful lens through which to view the science of attention. He invites me to join him and the magician for dinner, so I can see for myself.

"Distraction," Dr. Gazzaley says, "is a powerful weapon."

CHAPTER 8

TERRYL

ONE NIGHT, WHEN TERRYL Danielson was in the tenth grade, the door swung open in her bedroom. It was midnight, the cusp of another scary and vivid childhood memory. Her dad stood in the doorway again, this time armed with a saxophone.

He blew into it, his red cheeks filling it with breath and spittle, creating screeching noises, nothing approximating music. He was beyond drunk, blitzed, filled with fury.

"Stop it. We're trying to sleep!" she yelled.

"You shut your mouth!"

He began to parade around the house. Terryl shut her door and put on the record player to shut out the noise.

It was a new chapter with a similar theme and a more intense antagonist. At this point, Terryl recalls, Danny had started drinking earlier in the day, every day. He would wait until everyone was asleep, then take out Michael's gold saxophone, and start blowing, sometimes for hours.

Terryl would hide the saxophone. In the bathtub, a cupboard, her bedroom. Danny would wake her, demand it, berate her, threaten her. Her beatings for the most part had stopped. This was the new psychological warfare. Sometimes he'd find the sax. Sometimes she'd get the sleep her body craved, that she knew she needed, even though she wasn't trying to stand out at school, just survive it. Somehow she got A's and B's.

IN HER DIARY, SHE graded her days. Most were D's and F's. But there wasn't much to be done about it. Terryl remembers Danny making the stakes clear: If Kathie left, he'd take Mitchell. "The threat of him kidnapping Mitchell was constantly held over our heads," Terryl says. The cops were no help. Sometimes Terryl would call them, or a neighbor did, but she felt hopeless, and, like Danny, had it covered: He told the officers that it was a family matter.

Terryl also felt that Kathie was accusing her of being an instigator. Terryl thought she followed most of the rules: Don't mess up the house, stay outside, and don't answer the phone. She wasn't sure why she wasn't supposed to pick up the phone, but that was the rule.

Terryl felt like her mother blamed her for antagonizing Danny, and making things worse.

And there was the threat of gunplay. Her dad sometimes flashed his Magnum. Terryl remembers him also carrying a brown bag filled with painkillers, which he washed down with vodka.

Sometimes, when Danny was wasted, he would tell Terryl and Michael that he wasn't their real father. In the morning, Terryl would ask her mother about it and Kathie would tell her that Danny was drunk and to ignore him.

Terryl knew better than to invite any friends over to the house. But once, she invited the daughter of the bishop at her local church to visit her when Terryl was visiting her maternal grandparents. It was a sleepover, interrupted. In the middle of the night, Danny showed up hammered, and threatened to take Mitchell. Same old, same old, at least for Terryl. But she looked over and saw the bishop's daughter, Julie, tucked between the edge of the wall and the piano, sobbing. They took her home the next morning, a day earlier than they'd planned, and the girl never spoke to Terryl again.

TERRYL'S FIRST ESCAPE HATCH had been books. She took a more concrete step when she was sixteen. She went to a local youth employment program in Downey and got a referral to a mom-and-pop accounting business.

It was in a small office plaza in Downey, in a nondescript two-story office building owned by Mark and Millie Mandel. They were an older couple, both CPAs, on their second marriages. They had eight children between them—all grown, all college grads. Terryl earned $2.25 an hour answering phones and filing papers. She bought Mitchell a new pair of sneakers, sometimes a toy. She purchased a dirt-cheap, used red Ford Pinto with fake-wood paneling.

At the firm, she worked quietly, trying to be friendly but invisible, just like at Warren High School, where she tried not to fall asleep conspicuously. She reminisces that she managed a 3.2 GPA at Warren but that she felt like the "trashiest girl in school."

In her senior year, Terryl tried out for the cheerleading squad. One night, she went over to the house of Nanci Smith and her husband, Rich, the family's close friends from way back, and showed off the cheers she was working on. Terryl was so excited. Then Danny came over and was drunk, or on the pills, and just started ripping into Terryl, telling her she was nothing. "He slammed her verbally, up one side and down the other," Nanci remembers. "He crushed her like a little bug."

Nanci thought that maybe this big drunk went after Terryl because, in his own way, he was intimidated by her. "She could outthink him. She was smarter than him, always two steps ahead of him," Nanci recalls. It frustrated Danny that "she wouldn't cave in. He couldn't destroy her with words."

Nanci remembers that Terryl was the one in the family who would dump out her dad's booze. Not Kathie, because she knew what Danny would do if he found an empty bottle or cup. "The you-know-what would hit the fan," Nanci said Kathie would tell her.

Nanci believes Kathie didn't leave Danny because he did provide for the family financially. When Danny was sober, things were pretty good, Nanci remembers. "They never went without." Besides, "Kathie was good people but she didn't have much education to fall back on." It was a different time; wives just didn't up and leave husbands. Plus, Danny generally would never abuse anyone in public, not physically, at least, making it his word against the family's. The cheerleading incident was the rare time when he lashed out while people were watching. "He was very shrewd that way," Nanci says.

Others saw only mean. Another family friend, Patricia Dian Hauser, says Danny would be drunk by ten in the morning, yelling, threatening violence. "Kathie was a mess so many times at my house," Patricia recounts, and she remembers Danny using Mitchell as a pawn, threatening to take the boy, or refusing to let Kathie take him when she sought escape.

Michael, Terryl's older brother, had trouble going against Danny, Patricia recollects. Less so for Terryl. "Terryl fought back," Patricia says. "She was the fiercest little person."

For his part, Mitchell, Terryl's younger brother, remembers things differently, although he was a baby at the time. Looking back years later, he recalls his dad as a great guy. "My hero," he says of Danny. Yes, Mitchell says looking back, "there were problems," but he remembers them stemming from marital strife and tension between Kathie and his dad, not any cruelty on the part of his father.

In her senior year in high school, Terryl wound up making the cheerleading squad and got two rewards: a royal blue, gold, and white outfit, and, along with the library books she voraciously consumed, another escape route. She went to practices in the morning and after school. She was out of the house. She cheered for football and basketball players. She tried to be bubbly.

MIDWAY THROUGH TERRYL'S SENIOR year, she says her mom had finally had enough of Danny and went to Norwalk Superior Court to officially get a divorce. She took Terryl with her. Terryl dressed nicely, with a white blouse and a Peter Pan–type collar. She was proud of and pleasantly surprised by her mom's courage. But there was a much bigger surprise waiting in the courtroom.

Partway through the hearing, Terryl remembers, the judge asked Kathie: "Does the biological father pay child support for the minor daughter?"

He meant Terryl.

She says she looked at her mother: "What biological father? What biological father?!"

That was how she found out that Danny wasn't her real dad.

Terryl pestered her mother for details but got few. Terryl wasn't told her father's name. She learned only that he'd been a high school sweetheart who was long gone.

"He doesn't want you, and he's not interested in being a father," Terryl says Kathie told her.

Terryl felt relief that Danny wasn't her blood. He didn't totally define her. But she was angry, too. There was another man out there who was her father "who didn't care about me."

FOR ALL THAT, IT wasn't until high school graduation that something happened that hadn't occurred in years: Terryl cried. Sitting in her cap and gown, a sunny day, she let it go. It wasn't sadness, exactly. It was determination.

"I knew I never wanted to see any of those people again," she says, looking back. "It was one of the only times in my life that I've cried."

THE TRASH CAN HELD a pile of refuse, and a woman, upside down. Dead. Her blood-slick feet stuck out of the top of the garbage.

"What is this doing in an advertisement for shoes?" demanded Terryl.

She stood on a small stage at Cypress College, her second year there. On the screen behind her was the image of the dead woman. Not a crime scene, a shoe ad. Terryl was doing a command performance for her debate team, arguing about the deleterious impact of violence in advertising.

Terryl's high school guidance counselor didn't consider college an option, urging the young woman to go to a trade school. But Terryl found a community college, feeding off encouragement from the Mandels, the family she worked for at the accounting firm, and her own raw faith that she could change her circumstances.

She recalls she got further encouragement at Cypress from a communications professor named Pat Ganer. She was heavyset, dowdy, strict, seemingly old to Terryl, though she was only in her thirties, and a caring talent scout.

"Terryl, what would you think about joining the debate team?"

She posed the question not long after Terryl arrived at Cypress, a few months after high school graduation. The professor saw drive and strength in the young woman, the voracious reader with the discipline, Terryl recounts.

Terryl, starved for such encouragement, lapped it up. She stayed for hours in Dr. Ganer's office, getting help with homework. She spent even more time with the debate team and at tournaments, once arguing in the persuasive category in favor of child support. For the first time, she drew power from her personal experiences; she'd hidden and run from her home life. Now it was also serving as a source of drive, even inspiration.

She told Dr. Ganer a bit about Danny. The next year, the professor suggested Terryl do her persuasive address on violence in advertising. She leapt into it, finding numerous examples she'd show on the screen during her talks, like the one where the husband led the wife around by a leash, she on all fours like a dog. It was an ad for a department store.

TERRYL HAD TAKEN A job at Disneyland as a ride operator. She worked on Space Mountain, proudly wearing the uniform, a red, white, and turquoise polyester jumpsuit. She met lots of young people, many attending college at nearby USC. They joked in the break room, they laughed. As advertised, Disneyland, to Terryl, was the happiest place on earth.

Encouraged by Dr. Ganer and the young people she met from USC, Terryl applied after two years at Cypress to a bunch of colleges: George Washington and Georgetown, the University of Kentucky, UCLA, and USC. She got into every one of them, which, still highly uncertain of herself, she silently attributed to the fact that she was poor and filling someone's quota. But she was still elated, particularly when the envelope arrived from USC, which was attended by all those happy, privileged kids from Disneyland.

At USC, she took a class in criminal psychology, which she thought would fulfill a credit but wound up touching her deeply. She knew well how "bad men" acted, after years of living with Danny. She started thinking about becoming a prosecutor. She took classes at the university's Annenberg School for Communication. It was interesting stuff, but it also felt practical. She might not be able to afford law school, she thought, but she could always make a living doing public relations. She liked having a glass-half-full outlook and was adept at putting a good face on things.

She got scholarship money to live in the dorm. Her roommate was a music major, which meant Terryl got to attend music events. She joined the "Helenes," the smiling, effervescent greeters at USC sporting events. She told no one about her past, her home life, and they didn't seem to care. She felt safe. The dorm had a guard. She could sleep without keeping one eye open, as she had to do at home.

Danny, though, would not let her go. Far from it.

One late afternoon in the summer after her first year at USC—her junior year, given all her Cypress credits—she took Mitchell to the movies at the Cerritos mall. Mitchell was now six, and a sweet little guy. The pair left the theater through the back doors. By now it was dark. In a vivid recounting, Terryl says she didn't see Danny until it was too late: He was standing right next to them, wearing the familiar blue work pants and shirt, literally stinking drunk.

"Terryl, say good-bye to your little brother," she recollects him saying. She looked at him. What did he mean?

"You're never going to see him again!" Danny grabbed Mitchell by the arm.

"No! No! No! Help! Help!" Terryl screamed.

She grasped Mitchell by the other arm. She pulled, and Danny pulled. And then Danny, so much bigger than Terryl, swung Mitchell away.

"That's when I lost him. I didn't have a grip on him anymore. Danny grabbed him and ran away and threw him in the truck and they drove off."

Terryl remembers running to her Pinto after the confrontation with Danny. She climbed in. She became paralyzed. Tears poured from her, hysteria. It was the days before cell phones—she couldn't call her mom. She didn't know if she should call the police. What could they do? What had they ever done?

She sat in the car for a long time. Still hysterical, crying the way she hadn't in many years, maybe since high school graduation. In that state, she drove home, blind with anger and helplessness. It nearly cost her dearly. She was so consumed that, driving on the highway, too fast, she rounded a curve and the Pinto swerved; she lost control, and then, at the last second, recovered.

Danny returned Mitchell unharmed, but another scar was left on Terryl, the seeds of a recurring nightmare. Whenever she slept at home, she'd have it: her dad taking Mitchell, threatening that Terryl would never see the boy again. And, in the nightmare, she never did see her brother again.

Mitchell, looking back, remembers the incident differently. He says his dad came to get him because Terryl had him out too late. His dad, Mitchell recalls, never got out of the car, and only got upset when Terryl wouldn't hand him over and resisted. "She was always mouthie," Mitchell recalls, suggesting Terryl provoked Danny. And he says his dad took him right home that day to Kathie. His memory underscores some differences in how the children saw their parents, and the strife. But Mitchell does concede there were family problems, to the point that he says that when he was in third or fourth grade, after his parents split, he asked to live for a year with his babysitter "because of the drama."

"I kind of threw down the gauntlet as much as a little kid can, and told them 'I don't want to live with either one of you,'" he recalls.

For his part, Michael, Terryl's older brother, remembers the violence starting when he was five and Terryl was four. Danny "beat on me until I was eighteen, and then one night at the wrecking yard, he was drunk and started to get violent with me and I pushed him hard (I was a bit bigger by then) . . . he said he was going to get his gun . . . I could not allow that," Mitchell wrote in an email to Terryl, reflecting on their childhood. He wrote that he and Danny got into a terrible fight at the wrecking yard. In another email, he wrote of how Danny would tell him and Terryl he wasn't their father. "When he was saying these things he usually had me slammed up against a wall yelling in my face with his putrid booze breath."

ON THANKSGIVING DAY OF Terryl's senior year in college, she recalls that she and Kathie went for a walk on the beach. Kathie had news. Terryl's real father had made contact, along with his new wife. The pair and Kathie had dinner. Kathie showed them pictures of Terryl and Michael. Terryl says Kathie told her that the man was curious about how the kids turned out but didn't want to be a father to them.

Terryl was furious at her mother for showing her picture to a man who didn't want anything to do with her.

HOW WAS SHE GOING to escape this life? It followed her, even in her dreams. Was there a way for the bubbly and optimistic to overcome the terror and loneliness?

One thing was for sure: There would be no family for Terryl, no marriage. Earlier in her life, she'd told her diary that she was going to have a family and get married in the Mormon temple. But she'd changed her mind. She couldn't risk putting a child through what she had experienced. There was just too much she couldn't control.

CHAPTER 9

REGGIE

TROOPER RINDLISBACHER MADE PLANS to do follow-up interviews with Kaiserman the farrier, and with Reggie.

A few days after the accident, he called Mary Jane to set something up. It was around nine in the morning, right before she was going to go out and deliver the day's mail.

"I don't feel good about doing this without a lawyer," she told Rindlisbacher.

She remembers the conversation went sharply south. "He started to accuse me about knowing something," she says, looking back.

He asked if Reggie was texting at the time of the accident.

Mary Jane says she was offended by the substance of the accusation—and by the tone. Trooper Rindlisbacher was "awful," she says. "I've never been treated so nasty in my life."

She sensed a chip on his shoulder. *What's with this guy?* It rankled Mary Jane because she had such respect for law enforcement.

As to substance, she truly believed that the accident had been caused by the weather, hydroplaning. She got off the phone in tears, "scared to death."

Still, the two of them had managed to set a tentative plan for a follow-up interview with Reggie.

A few days later, Rindlisbacher drove to the Tremonton City offices and met with Kaiserman. The farrier went through detailed drawings of what had happened. He was upset, pained at having been part of the tragedy.

Then it was time to meet with Reggie. Rindlisbacher's phone rang. It was an attorney hired by the Shaw family. The lawyer told him that Reggie wouldn't be coming for an interview and that he, the lawyer, would answer all questions.

Rindlisbacher was irritated. He remembers thinking of Reggie: "He's lawyered up, and he's not going to say anything. He's got something to hide."

MARY JANE OPENED THE bedroom door. Reggie was right there on the bed, where she had left him, facing the wall, his phone on the bed, behind him.

"Reggie."

He half turned. It had been another restless night. He kept replaying the accident, thinking about the two dead men and their families, worrying what might happen next. His mom sat on the bed.

"I think you should go talk to someone."

He knew what she had in mind. A counselor named Gaylyn White had an office just a few blocks away, behind the office of Russell White, Gaylin's husband, Reggie's dentist and a local church leader.

"No."

"Just to get things off your chest, Reg."

Something was obviously wrong. Two days after the wreck and Reggie had barely left his room; hadn't left the house at all. And she didn't know the extent of his emotions. "I felt like I didn't deserve to go out there and live," he says. "Not in a suicidal sense. In the sense that I didn't deserve to enjoy my life."

Not when those two men in the other car had no life to enjoy.

Mary Jane left. Reggie thought: Crazy people go get help. Not sane people.

Not guys' guys. Not athletes, not in Tremonton.

"IF YOU'VE SEEN *FRIDAY Night Lights*, that was us. Same kind of place, same kind of problems," says Dallas Miller, Reggie's best friend. The jocks held sway, but the focus on sports, the all-American veneer, had a dark underbelly, at least from Dallas's perspective. A lot of drinking—weekend nights downing Keystone beer in the foothills—sometimes driving home drunk—a lot of premarital sex and a lot of pregnancy. "Not to put a bad rap on our town, but, more often than not, not everyone on the sports team was living their religion."

It was far from everyone's view. In Dallas's case, he felt he was in a better position to see past the town's facade. A rebel from his earliest days, he rejected the Church, thinking it a place for hypocrites who didn't live the life they preached. At odds with his parents, he spent many nights on Reggie's floor, sometimes passed out from too much drinking.

In fact, there was a decent chance he could've been the one who wound up involved in something tragic. But Reggie? When Dallas heard about it, he thought: "He was the last guy I'd have expected this to happen to.

"He was the kid who always made the right choices, who always did the right thing."

Reggie could cool Dallas off. In January of their senior year in high school, Dallas, six foot two and 205 pounds, threw the ball in someone's face during basketball practice, hitting his teammate in the head. The two squared off and a fistfight seemed imminent. "Reggie grabbed me and threw me in the locker room. We had a long talk," Dallas recalls. He adds: "Reggie was the only person I took advice from.

"He was a listener, and after he did the listening part he would have something to say that was probably in my best interest."

Certainly not the kind of person who would make a bad decision and kill two men. It must've just been a horrible accident, Dallas figured, no one in the wrong, but a tragedy nonetheless.

"I was the wild one. He was the mild one. The good listener."

REGGIE'S FIRST AND DEEPEST rival was his older brother Nick. People thought Reggie and Nick were twins; Reggie was big for his age, and Nick small. Sometimes they thought Reggie was actually older. They competed at everything, stoked by a culture of competition. "I used to make them fight for crackers when I babysat," says older brother Phill, mostly joking. "Only the winner got to eat."

They played basketball and baseball outside and football video games in the house—sitting on the floor or couch, with controllers in their hands.

They got their first console, a Nintendo Entertainment System, for Christmas in the early 1990s. It was a square gray box with rectangular controllers that featured a few simple buttons. They went at it in Tecmo Super Bowl, a football game. Reggie was little more than five years old. Several years later, the brothers moved on to a different console, the Sega Genesis. It used a 16-bit processor, which allowed double the processing capability of the 8-bit Nintendo, an early sign that the power of these machines would soar with each generation. Better graphics and sound, more complex challenges and on-screen data to juggle.

They graduated to Super Nintendo, also a 16-bit system, and Mario Kart, a racing game. A friend of Phill's named Ryan was living in a one-bedroom apartment located behind the Shaws' house. On summer mornings, Ryan would knock on the back door and get Nick and Reggie out of bed to play Mario.

"We'd play for hours and hours. And if we had baseball or football, we'd take a break and go practice and then come home and play again," Reggie remembers.

When it came to technology, Reggie had come of age in an extraordinary time. Just a few years before he was born, in 1983, *Time* magazine did a twist on its person of the year feature and instead named the personal computer its "Machine of the Year." The cover heralded "The Computer Moves In."

REGGIE'S OWN ADOPTION OF technology very much paralleled what was happening around the country, a phenomenon documented in a series of pioneering studies initiated by the Kaiser Family Foundation in 1999. They show an explosion in use of media by young people.

In the first study (1999), Kaiser found that the average child spent five hours and twenty-nine minutes using media per day. Within the average, older children ages eight to eighteen spent nearly double the time with media than toddlers. Boys spent slightly more time than girls, and minorities slightly more than whites.

In 1999, television was the media that got most of the attention, at about two hours and forty-five minutes per day, compared to eighty-five minutes listening to music (tapes, CDs, or the radio), and forty-four minutes reading or being read to. At that point, only about twenty-one minutes were spent using the computer for fun, and twenty minutes on video games. There was virtually no Internet use.

The second study, in 2004, showed big change. There was massive access to new kinds of media that hadn't existed in the mainstream just five years earlier. Take instant messaging, which the 2004 report found to be in 60 percent of homes with children, but that hadn't been in virtually any home five years earlier. That second report found that 74 percent of homes had Internet, up from 47 percent in 1999.

The 2004 report found that about 40 percent of children age eight to eighteen had a cell phone (most on the upper end of that age range). There wasn't even a baseline from the previous study, given the low use of cell phones at the turn of the century.

The third report, in 2009, showed another leap in media use by young people. The average eight- to eighteen-year-old was consuming ten hours and forty-five minutes of media per day.

How? How was it possible to use almost as much media as there were waking hours? The answer: multitasking. The study found that the typical person in that age group spent about 7.5 hours using "entertainment media," but that a good chunk of that was spent multitasking— using more than one source of media at a time—and those hours were double-counted by the researchers.

Their attention was divided a good chunk of the day even within the media space.

The researchers found the change owed in large part to access by children of personal devices, like iPods and cell phones; about 66 percent of eight- to eighteen-year-olds had cell phones, not quite double the percent from the study five years earlier.

Of note, the new media appeared to be taking time from the biggest screen draw of them all—television. The study found that children's viewing of traditional television had fallen twenty-five minutes per day from 2004. But a look deeper shows a catch. Kids were now accessing television programming on computers—including phones, the Internet, and other gadgets. As a result, overall consumption of television content actually grew. By now, children were watching 4.5 hours of television *programming* a day, up from three hours and fifty minutes in the previous study. And by 2009, roughly an hour of video consumption was done on a device other than the TV.

There was a fascinating omission from the 2009 report. It did not include texting or talking on the cell phone in its definition of media use. That was extra. On average, the report found, the average high schoolers texted for an hour and thirty-five minutes each day. They were doing that in addition to the other media and tasks.

They, and others, were doing it while driving. In 2007, a survey by Nationwide Insurance found that 73 percent of people reported talking on a cell phone while driving, with teen drivers showing the highest use of any demographic.

While Reggie came of age, teen media use soared, and multitasking with it.

REGGIE DIDN'T NEED TECHNOLOGY to be social. He was plenty charismatic, but in a quiet way. He didn't demand attention but everyone liked him, including the girls. He spent a lot of his waking hours thinking about them. The first one he really lost it for was Cammi. He met her through a friend and the chemistry was powerful.

His friends and family weren't so sure about the whole thing. At first, for instance, Dallas didn't like Cammi. It bummed Dallas out that Reggie spent less time with him than with this tall brunette, who Dallas thought was just okay, and reasonably pretty. But Reggie and Cammi would be locked at the hip, and the lips. Some light public smooching, sitting on the couch holding hands.

Then Dallas came to appreciate the fact Reggie was so into someone who seemed so into him. "I was a jerk to her at first. But she was a real sweet girl."

Reggie's mom felt otherwise.

"Everyone in town knew she didn't like Cammi," Dallas says of Mary Jane. "She didn't want her son being serious and being stuck in town."

Mary Jane makes no bones about it. "She was darling to look at, but she had no personality whatsoever," Mary Jane says, pulling no punches. "She was awful."

The reason for this strong reaction? Mary Jane believed Cammi was a roadblock to Reggie going on a mission. To Mary Jane, the young woman was putting her passions and wants ahead of Reggie's dream.

Valid or not, Mary Jane's conspiracy theories underscored how invested the family was in Reggie's mission, something he professed to want so badly himself.

After high school graduation, Reggie went to a year of college in Virginia, and to play basketball. He was good enough to make the team but not good enough to get a scholarship. Ed doubted if Reggie knew it, but the family paid the tuition in no small part because they wanted to get him away from Cammi.

Yes, the family wanted Reggie to fall in love and get married and have a family. But the mission came first and everything would follow.

AT VIRGINIA, REGGIE HAD a good time, and was a decent student. He developed some confidence and a better jump shot.

In May, the semester over, he came home to go on his mission. In June, just a few months before the accident, Lisa and Van Park attended his farewell at the Garland Tabernacle, a modest red-brick church with spires on each side. Inside, dark wood pews beneath cylindrical lights hung from the ceiling. Reggie gave a brief talk. Afterward, the group went back to the Shaws' house and had a quick celebratory lunch.

Then later that day, Reggie went over to visit the Parks at their two-story, five-bedroom house just a block from the high school. They sat outside and the Parks gave their well-wishes to a boy who they imagined was the kind of kid they wouldn't mind seeing their daughter with.

"I was so excited for him to go to Canada," Lisa reflects.

Just a few days later, Reggie returned home, ashamed.

"Knowing him, I'd have thought: He's just got to clear some things up. He'll get it together and he'll just go back," Lisa says. "I never felt 'What a disappointment.' If you're Christian in your heart, that's not how you act."

Then September 22 came around. That night, Lisa heard about the wreck from Van. But the intensity and the severity of things didn't quite hit until a few days later, when she paid her regular visit to a small nail salon in Tremonton. The woman who gave Lisa her clear polish and French tip was Chantel Gubeli, who was married to the brother of Cammi.

"Can you believe what happened to Reggie?" Lisa recalls Chantel asking her.

The more Chantel told Lisa about the accident, the more Lisa's heart broke. It must've been the road conditions, or some freak accident. It couldn't have been Reggie's fault. "He's more mature than that. He's more responsible than that.

"She told me about the families of the men who'd been killed," Lisa says. "I remember thinking: 'There may not be a chance for them to forgive him.'"

CHAPTER 10

REGGIE

MIDMORNING, THREE DAYS AFTER the accident, Reggie walked out of the house for the first time. It had just snowed. Reggie thought: *It is so bright out here.* He put the keys into the ignition of the Chevy. He drove alone. Every car that went by was a terror, a potential fatality. Something just clicked. He'd see the counselor after all.

Up the stairs, he found the office of Gaylyn White, the counselor his mother had asked him to see. Reggie had relented. The office was small, with a floor-to-ceiling bookshelf along one wall. Gaylyn sat behind her desk, a simple thing with a laptop on it. He took a seat in one of the chairs in front of the desk. Gaylyn noticed in particular Reggie's sad eyes. She'd known him a long time, and he always had sad-looking eyes, she thought, even though he was an easygoing, happy person. Now there was nothing playful beneath the sad eyes.

She knew about the accident, of course. She wanted to hear his telling of it. He started by telling her about the morning of the accident, how his mom had gotten him out of bed because he'd slept through his alarm. By then he was crying, sobbing, pulling at the box of Kleenex on Gaylyn's desk. It was bothering him, killing him: He knew the story's terrible ending, of course, but a lot of the details were missing. He couldn't picture what happened.

He told her about driving the SUV over a hill and then going down the other side. The weather was bad.

"I crossed the center line, just a little bit, and I hit the car."

He was sobbing.

"Reggie, why do you think you can't remember what happened?"

He shrugged. He thought: *Because the details aren't important to me. What really matters is that those two men are dead and what are their families going to do now?*

Gaylyn listened and typed notes into her laptop. She tried, as always, to keep her typing unobtrusive, so Reggie wouldn't get distracted. But she was typing even less than usual. That's because she realized early on in the conversation that she ought to keep the notes modest, in case there were legal implications should Reggie eventually face charges. Even recognizing there was doctor-patient confidentiality, she wanted to be careful about what she put into words.

She asked Reggie why he thought he was having trouble remembering what happened at the time of the accident. He didn't know. She posited: "You probably went into shock. That's why you're not remembering."

Reggie nodded. It sounded right. "I can't believe I could do something like this."

He couldn't remember. It had just been an accident, right? So why did he feel like he'd done something so wrong?

Gaylyn gave Reggie some tasks: exercise, write in a journal, meditate. She said they should meet again, and he agreed. She told him he should start writing letters to the families of Keith and Jim, not necessarily to send them but just to express himself.

And so he went back to his room and started writing. Letters and letters to the families of Jim and Keith. They read, over and over again: *I'm sorry. I'm sorry. I'm sorry. I'm sorry.*

AFTER THE SESSION, GAYLYN couldn't help but think about the larger picture of Reggie's life. He'd just returned, under difficult circumstances, from his mission. Not that she judged him. But she also understood that Reggie might not feel that way, that he might be carrying a heavy weight of feeling like he'd let down himself and his family and his community. "I can't imagine having to come home," she says of the idea of returning from a mission early, and under a cloud. She says she wonders whether people who came home prematurely might be better off living somewhere else for a while, so they would not have to deal with feeling judged. "I'm amazed people come home to the community. I don't know how you go home to the community."

TEN DAYS LATER, ON October 5, Reggie returned to Gaylyn. This time, he told her he was feeling better. "He was acting like he was doing okay," she recalls. He exhibited less anxiety. Her notes, released with Reggie's permission, read: "He reports some ruminating and wishing more than anything else he had not been on the road at that moment."

Gaylyn wasn't particularly buying that Reggie's mind had calmed or his anxiety and even depression had lifted. She suspected that Reggie was aware that the $65-an-hour fee she charged—there was no insurance for her visits—could be a hardship on his family. Reggie, she thought, hated to displease people, particularly his family. There were so many pressures on him: not to let his family down, particularly after the mission return, and not to compound whatever shame he perceived they felt by him getting in trouble with the law, going to jail, or whatever might happen.

Gaylyn talked to him about understanding that the past is a path that is behind us, not the path we are destined to take in the future.

Of course, she felt strongly she should see him again but doubted she would.

She let her mind connect dots between the accident and his failed mission. She wasn't sure of the details, but she could surmise that he'd told someone at the Mission Training Center in Provo that he'd done something inappropriate in the lead-up to the mission.

That much was commendable, she thought. Ultimately, he had told the truth. But if he'd gotten to the training center in the first place, it surely meant that somewhere along the line he probably lied. He'd had to have told someone he didn't have any problems, such as premarital sex, to bar his place on the mission.

And so Gaylyn was left wondering whether Reggie was sincerely traumatized by the accident, and therefore couldn't remember what had happened, or whether there was something more insidious occurring.

"I was wondering whether or not his sense of conflict about the accident was genuine confusion, or another lie."

SHE WAS RIGHT ABOUT the mission. Reggie had lied, several times, in the lead-up to his mission. He had lied to his bishop about having sex.

His lies weren't particularly big or unusual. After all, it's not uncommon for young people to lie to parents or a pastor about their having premarital sex. And they tell those lies even when there isn't the intense cultural pressure that Reggie felt as part of a small Mormon community. The very language used by people in Tremonton betrayed just how intense that pressure was. Mary Jane feared a "disgrace" when Reggie came home. Gaylyn even wondered about how people come home after a failed mission. The environment explains how Reggie could have felt himself in a fishbowl or crucible; how he'd put his family in a bad light.

What Gaylyn was wondering as Reggie sat in her office was whether the first failed mission had upped the stakes for Reggie. He'd already felt he let his family down once. Now, so soon after coming home, he was behind the wheel during a deadly wreck. What if he'd done something wrong? How could this young man sustain another disappointment?

And Gaylyn didn't know the full story—just how the lies tortured Reggie and left him questioning how he behaved when forced to confront difficult truths.

DURING THANKSGIVING BREAK OF Reggie's freshman year in Virginia, he came back to Tremonton to see the family, Cammi, and his bishop, David Lasley, who lived just down the street from the Shaws.

The bishop is an important job, but perhaps not as significant as the term *bishop* might communicate to people of other faiths. The bishop is a lay leader for the local ward, which is one of the small but critical parts of the highly organized and hierarchal Mormon structure. Each ward has about five hundred people, and roughly ten wards make up a stake, which has a president.

Reggie set up a meeting to talk to Bishop Lasley about going on a mission, Reggie's lifelong dream. In a modest office, the bishop asked Reggie about his interests, passions, and intentions. He also asked if Reggie had violated any tenets, like having recent premarital sex. Reggie assured him that he had not.

It was the truth. But not for long.

A few weeks later, Reggie came home again for the Christmas break. One afternoon, he found himself at Cammi's house. They'd been talking about this moment lately. They went to her bedroom, her queen-sized bed.

She told him she loved him. He told her that he loved her, too, the first time he'd said that to her, to anyone.

They did it. It was good, it was fun. "I was nineteen, and there's only one thing that really seemed important." Love or sex? "Probably sex," Reggie says, looking back. "I thought it was love, but thinking back, lust and love, they seemed awfully close to each other at the time."

And commingled with them was guilt. A few days after he lost his virginity, he went, as always, to church. He couldn't pray.

"I tried my best to act like it was a normal Sunday and go into church, but it didn't feel right. I didn't belong."

A few days later, Reggie went back to the bishop's office for a follow-up meeting.

"Things still good on the girl front?" Bishop Lasley asked.

"We're good. Everything's good," Reggie said. He tried to look the bishop right in the eye.

A few hours later, he saw Cammi. She asked: "Did you tell him?"

"Yeah, yeah, I told him." He explained to Cammi that the bishop said Reggie could get a onetime exception. "He thinks that I'll be okay and I'll be able to go, as long as we're good and as long as we've stopped."

He'd lied to Cammi, too.

He was afraid that if he told her the truth, she might tell the bishop, or someone, thus ensuring he would stay home and be with her, and not go on a mission.

"She didn't want me to go. She wanted me to stay home and get married."

Back in the office with Gaylyn, the counselor, the issue didn't come up. But it was weighing heavily on Reggie, and hadn't stopped weighing on him for many months. He felt in some way that it was a kind of original sin, a terrible thing he'd done that spoke to a dark thing inside of him.

"It was a big point in my life, and I chose to look people square in the eye and lie about it."

IN BETWEEN HIS MEETINGS with Gaylyn, on September 28, Reggie left the house a second time. He had a midday appointment to see Jon Bunderson, a lawyer who Reggie's dad, Ed, had gotten a referral for. Bunderson had an office twenty minutes away, in Brigham City, where Ed worked in a machine shop.

Mary Jane drove Reggie from Tremonton, and they picked up Reggie's dad at the Brigham Implement Company. Ed's employer sold farm equipment but also had a machine shop that he had managed for nearly a decade, helping building it up into a substantial operation that did metal fabrication, built parts for airplane Jetways and automobile airbags, and took on odd jobs, like fixing a bicycle or making parts for an irrigation ditch.

It was blue-collar all the way, located inside a corrugated metal shell; metal shavings covered the floor around the heavy equipment, which filled the shop with a regular hum. Out front was a line of loaders and tractors waiting to be repaired. But you could hardly beat the shop's setting—if you looked west, beyond nearby Brigham City, the Rockies exploded dramatically toward the sky.

The family drove to Bunderson's office. It was located in a single-story building near the county courthouse. It was unassuming to the point of being ugly, with a pink, unkempt exterior and a mass-manufactured mailbox with an eagle on top of it. It cut a stark contrast to the impressive municipal offices and other tall, well-manicured downtown buildings of this historic city, where, in 1877, Brigham Young delivered his final sermon.

Inside Bunderson's office was a reception area covered in green-striped wallpaper. Despite the less-than-ornate surroundings, Reggie immediately felt he could trust Mr. Bunderson. Reggie thought of him as a smaller man physically, in his late fifties, with a mustache, wearing a suit and tie. He had a quiet intensity, a bit of gravel in his voice. He seemed authoritative and smart. He asked Reggie the basic questions about what happened the morning of the accident. And Reggie replayed what he remembered, tearful but not as emotional as when he'd met the therapist. When he faltered, his parents filled in the blanks.

Reggie concluded by saying he thought the accident was caused by the wet roads.

"I remember the weather that day," Mr. Bunderson concurred in the meeting. He said he wasn't surprised that something tragic could result from the rough conditions.

He asked Reggie if he'd been on the phone during the accident, not just before it. No, Reggie told him.

Bunderson asked if he might get a copy of the phone bill. And Reggie's family said: Sure.

Bunderson gave Reggie assurances, and a few instructions. He told him he should not make contact with the families of Jim Furfaro and Keith O'Dell. Even though Reggie hadn't done anything wrong, "apologizing was admitting guilt."

"Things should be fine, Reggie," Bunderson told him, and addressed the whole family. "I'll call you if I hear anything."

Reggie felt some relief, reassured that jail wasn't imminent. And he felt he'd unburdened himself about the accident. As he looks back, he says: "I continued to believe what I was telling him was everything I knew and everything I remembered."

ON OCTOBER 21, A few weeks after the meeting, a seventeen-year-old named Michael "Taylor" Wick was driving south on Highway 91, in Cache County. His passenger was Christopher Lee Dorius, also seventeen. Both were football players at Sky View High School in Smithfield, the second largest city in Cache County after Logan. It was 6:15 p.m.

According to the highway patrol, Wick drifted off the road, overcorrected to get back onto the road, and was struck at high speed by a pickup coming from the other direction. Both young men were killed on impact.

The preliminary report could find no cause for this seemingly inexplicable accident. It just seemed like another teenage driving tragedy.

"He's been a part of the team for four years—a major part of the team," Sky View coach Craig Anhder said of Wick. "It's just a shock."

Eventually, there was a possible explanation. The local prosecutor's office was told that Wick was holding his phone at the time of the accident. But prosecutors didn't pursue the case because the two boys were already dead.

CHAPTER 11

THE NEUROSCIENTISTS

I N THE NINETEENTH CENTURY, Helmholtz and Donders had
begun trying to measure the capacities of the human brain. The next
wave of pioneering brain scientists, a remarkable group of men and
women, took their inspiration from World War II and the struggle of
human operators to keep up with the staggeringly powerful machines
and weapons of war.

One of these pioneers was Anne Taylor, who was a little more
than five years old when World War II started and the bombs rained
down. She'd hear the piercing siren. Her mother or father would scoop
her up along with her younger sister, Janet, and take them down to the
damp and musty cellar. They'd also take Tinkle, the family's tabby cat.
And they'd wait. Sometimes they'd take their pencils and scrap paper
and draw to pass the time.

As she got a bit older, still enduring the Luftwaffe's bombs, her
drawings showed curious little doodles. They represented the German's
deadly planes flying overhead.

When the "all clear" message would come over the loudspeakers, the
family would ascend to the ground floor of their little house in Kent,
England, midway between London and the Nazi airstrips in France. Her

mother was a homemaker. Her dad was a chief education officer for the Medway towns of Rochester, Gillingham, and Chatham. In the living room, there was a map on the wall with pins in it, and Anne's dad, Percy, would move the little flag pins to show what was happening at the front.

Around Britain and across the world, war raged. It was a mechanized affair. Men in planes and tanks, or armed with guns, artillery, and advanced weapons, tore apart the bodies of millions of soldiers and civilians. At the same time, not unrelated, machines were being put to important scientific use: Researchers were trying to measure the ability of pilots and soldiers to sustain their focus while operating the advanced weaponry.

How could pilots navigate these powerful planes—traveling at hundreds of miles an hour—while looking at the cockpit gauges, listening to the radio, evading antiaircraft fire, and dogfighting? How could soldiers on the ground, bombs falling all around, call in the right coordinates for air strikes? How could the air traffic controllers keep track of the blips on the radar screen amid heavy fighting?

"People were looking at these screens, these very primitive displays, and looking for signals, like German planes coming overhead, and they often missed them," explains Alan Mackworth, a professor of computer science at the University of British Columbia, who also holds the title of Canada Research Chair in Computational Intelligence.

Mackworth wasn't born until 1945. But he well knows the science, from his own study, and because his father was smack in the middle of it. Norman Humphrey Mackworth, known as "Mack," was working for the RAF, the British air force. They weren't doing theoretical work; they were trying to save lives by helping pilots and radar operators stay alert and capable in the face of an onslaught of information.

"There was a huge crisis," says Alan. If you misread the radar screen, got distracted, fell asleep, well, people died. Villages burned. Without being too hyperbolic: Battles were won and lost, and wars, too.

There was a young man, Donald Broadbent, who volunteered for the RAF at age seventeen. He would later go on to do some of the world's pioneering work in attention science, alongside the senior Mackworth.

Broadbent's interest also stemmed from the war, and the basic day-to-day challenges he observed pilots facing as they tried to stay focused. The *New World Encyclopedia* quotes a former colleague of Broadbent's recounting an anecdote:

"The AT6 planes had two identical levers under the seat, one to pull up the flaps and one to pull up the wheels. Donald told of the monotonous regularity with which his colleagues would pull the wrong lever while taking off and crash land an expensive aeroplane in the middle of a field."

Paul Atchley, a professor of psychology at the University of Kansas, points to this collision course of man and machine during World War II as a central reason why scientists developed a new urgency around understanding the human brain.

"We had these highly motivated individuals—radar operators and pilots—who would miss attacks or drop bombs on the wrong cities. Why did they fail?"

They were running up against the limits of their own brains. "Technology was outstripping cognitive capacity," Dr. Atchley explains. "We could quantify the machine, but not the human. That's where cognitive neuroscience really started."

FOR HIS PART, NORMAN Mackworth had been born in India, the son of a British eye surgeon who performed cataract surgery for locals. The family then moved back to Britain, where Mack grew up in Aberdeen, and became a scientist, a tinkerer, a great piano player, with a bit of a short attention span himself. He was prone to moving from concept to concept, but making big strides in a handful of key scientific camps. So maybe, given his own impulse to jump around, it's not surprising that, during the war, he came up with the Mackworth Clock.

It was a black box with a point on it that turned in a circle. The point moved at regular intervals of a second. Except that sometimes it would skip the one-second interval and move after a two-second interval. Such jumps happened at periodic, unpredictable times. It was the job of an experiment subject to press a button whenever there was a two-second jump. Simple, right?

After about thirty seconds, the subject's ability to focus—his "signal detection"—went down markedly. No wonder the radar operators, sitting eight hours a day in front of screens, could miss these life-threatening blips, these German bombers that could kill their friends and family. But why?

Unknown to little Anne Taylor, doodling to distract herself while her family tried to survive the bombs, defended by the RAF, this intersection of man and machine—the struggle to survive in wartime—was setting the stage for the next wave of attention study. What had begun with Helmholtz and Donders was taking a next major turn. And Anne, who later married and took the surname Treisman, was destined to be one of the researchers at the center of that stage.

"ANNE TREISMAN IS BRILLIANT," says Dr. Gazzaley. "She was a pioneer."

Dr. Gazzaley sits at Maverick's, an upscale eatery serving American comfort food. The restaurant is a one-minute walk from the Gazzloft, so close that Maverick's lets Dr. Gazzaley and his girlfriend, Jo, take home the plates when they order out. They're regulars here, it goes without saying, and Dr. Gazzaley, as much as he craves new experience and new stimulation, cannot help but always order the same thing: the southern-fried chicken with a biscuit and greens.

Dr. Gazzaley is nothing if not a convincing figure. He's sold five of the other six diners on the fried chicken. Patrick Martin, the big-time magician, is among them. He's got a thick build, short curly hair, wears a leather jacket, and seems to be a careful observer with a mischievous twinkle in his eye.

He promises, later in the night, maybe back at the Gazzloft when First Friday starts, to show a trick or two—to demonstrate more about the power of attention and distraction.

For now, Dr. Gazzaley is talking a bit about how attention science unfolded, and the leading researchers upon which modern work is built.

What was so amazing about Dr. Treisman?

"She was crucial in helping us understand bottom-up attention," says Dr. Gazzaley.

ANNE'S FAMILY SURVIVED THE war intact. She performed well academically; so well, in fact, that she was among just a few students from her grammar school fortunate enough to get admitted into Cambridge and Oxford. Early in life, she'd expressed an interest in the sciences, but her father thought she'd have no culture and so at Cambridge, she studied French literature.

Her success earned her an offer for a graduate fellowship, but she thought spending three years studying a single medieval poet sounded restricting. She asked Cambridge whether she might pursue a degree in psychology, which was growing in prominence and credibility.

"They said in horror: 'It's all about rats!' And I said that might be interesting."

Much of the focus in psychology at the time was around behaviorism. B. F. Skinner, John Watson, and others were focusing on the idea that human behavior could be understood by how people reacted to things; they put less focus on what was happening *inside* the brain, which seemed at the time like an inscrutable black box, the contents of which could not be observed.

Dr. Treisman was inspired to think differently by one of her instructors, Richard Gregory. He did all kinds of odd and even amusing experiments, like trying to show how vision could be impacted through neck strain, something he demonstrated by walking around the classroom wearing a heavy helmet to stress his neck muscles. Then he'd see how his vision changed. Could the physical environment change what was happening inside the head? The work was a distant echo of the work of Donders and Helmholtz, and the idea that brain activity could be measured.

By now, Norman Mackworth was the first director of the applied psychology unit at Cambridge. Working with him was Broadbent, who at seventeen had observed pilot error in the RAF, and who was on his way to becoming a godfather of cognitive psychology. After the war, he studied auditory channels, trying to understand what we focus on, and how much information we can absorb and process and under what circumstances. It was the precursor to the cocktail party effect, named by another British researcher, Edward Colin Cherry, in 1953.

Another researcher, a Canadian named Donald Hebb, was exploring a different angle. He theorized that the organization of the central nervous system and neural networks were involved with and impacted attention. His name and work would become increasingly appreciated as neuroscientists delved beyond behavioral studies and looked at the physical structures of the brain. Hebb's seminal 1949 book, *The Organization of Behavior*, illuminated a new pathway for studying the anatomy of the brain's attention networks.

These were among the seminal researchers in the field, but there were others and, collectively, with the war behind them, they had a new luxury: Gone was the immediate, life-and-death pressure of figuring out how to help pilots and radar operators sustain attention under duress. And there wasn't really a thought that technology—in a general sense—could distract people on an everyday basis.

AFTER ALL, THE DAILY dose of computers or telecommunications was still far away from most people. In 1945, for instance, AT&T began introducing a kind-of mobile phone service, derived from military radio technology, in a few metropolitan markets in the United States. Far from being ubiquitous, it could only be used simultaneously by no more than twenty people in a single city.

In 1950, about 9 percent of American homes had televisions, the first screen that would become part of most people's lives (by 1962, 90 percent of households would have one).

Researchers, absent the urgent need to know why radar operators fell asleep, started a new generation of more formal, deliberate experiments, trying to measure the capabilities of the human mind.

"Behavioral psychology was turning into the cognitive revolution," says Dr. Treisman. How does the brain process information? How much. And, she says: "What kind of things overload the brain?"

IN 1957, TREISMAN MOVED to Oxford to pursue a PhD. She posted a flyer around campus, asking students if they'd like to be subjects in a psychology experiment, and got plenty of takers. A subject would show up in a small room with a short wooden desk adorned simply with a Brenell Mark 5 two-channel tape recorder and a pair of headphones. The subject would don the headphones and discover that each channel was playing a different passage from a book (the same book—typically, *Lord Jim* by Joseph Conrad). Dr. Treisman instructed the subject to listen to the passage coming into one ear—say, the left ear—and then to immediately repeat what he or she heard. The trick was to ignore what was coming in on the right ear.

Then Dr. Treisman added a twist. After the first fifty words or so of the recording, she switched the passages, such that what had been coming through the left ear was now coming through the right, and vice versa. About 6 percent of the time, as the subjects were repeating what they heard in the left ear, a word from the right ear, the unattended one, would slip through into their report.

To Dr. Treisman, there were two basic conclusions: first, "the attention filter is very effective," but second, "our filter is not a total block."

That was a relatively powerful discovery for the time. Broadbent, for one, had assumed the filter was total. He assumed people could focus on what they wanted and block out the rest. For all his contributions, Broadbent's assumption turned out to be simplistic in much the same way that, before Helmholtz, people thought reaction time was immediate, infinite.

So what kinds of things get through the filter? What interruptions and stimuli rise to the surface even when we're intent on focusing on something else?

In a nearby office, a colleague of Dr. Treisman named Neville Moray made an interesting discovery also by experimenting with subjects hearing two different messages through headphones (known as the "shadowing technique"). He found that subjects listening to one message would, in only about 30 percent of cases, recognize that their own name had been said in the unattended ear, or be thrown off, derailed, by the sound of their name.

Dr. Treisman herself made an indirect discovery about the kinds of things that pop to the surface, even when we're trying to ignore them. It happened by accident. At one point, she tried her experiment by playing passages not from *Lord Jim* but from *Doctor Zhivago* by Boris Pasternak, the book that launched the epic film. One day, a subject started doing something Dr. Treisman had never seen before: He started remarking about the passages from *Doctor Zhivago* that were being read in both ears. It appeared, on its face, to be a highly unusual capacity to attend to both streams of information.

"It turns out that he was Pasternak's nephew," Dr. Treisman recalls, laughing. The subject professed great familiarity with his uncle's work. She surmised that even when we are trying to actively ignore something, we might be paying attention to it somewhere in the recesses.

"More monitoring may be going on than we realize," Dr. Treisman says. "Even this reduced, unintended message will trigger recognition if the subject is important, relevant, or highly probable."

DR. TREISMAN'S WORK VERIFIED the limits of the attention filter. It also provided a key building block in one of the most crucial, emerging principles of the science of attention: There is a tension going on inside the brain. It is a tug-of-war between two different aspects of the attention system, one called "bottom-up attention" and the other called "top-down attention."

Top-down attention is what we use to direct our focus, say, on a work project, or listening to *Lord Jim* through headphones, or when driving on the road. Top-down attention allows us to set our objectives and focus on them.

Bottom-up attention is different. It is what allows our attention to be captured instantly, without our control, say, by the sound of our name, or a bird flying by, or the ring of the phone. Bottom-up attention operates unconsciously, automatically, driven by sensory stimulus and contextual cues.

Top-down and bottom-up attentions are both essential to survival, and so is the balance between them. If we had no top-down attention, we couldn't direct our focus on important goals. But without bottom-up, we wouldn't be alerted to new stimuli, including danger. Imagine a caveman being so focused on building a fire that he never heard the lion coming through the bushes.

Before Dr. Treisman and her peers, there was a sense that people could control the direction of their attention, even if their reaction times weren't instantaneous. What Dr. Treisman and her peers began to realize and define was a powerful clash going on inside the brain.

BACK AT THE GAZZLOFT, another First Friday is about to begin. And so is Patrick, the magician. He asks if he may borrow some money from the audience. A woman produces a $20 bill which Patrick examines with approval and just a hint of the magician's sense of drama. He stretches the bill, snaps it a couple of times, showing that it is, in fact, just a standard Andrew Jackson, nothing fancy, no tricks.

He stands beneath the tree that bisects the middle of the loft. Outside, neon light from the neighborhood leaks in from the big window. Some new music plays on Dr. Gazzaley's many speakers. But for the five partygoers who are earliest to arrive, all eyes are on Patrick. Specifically, on his hands.

With his right thumb and finger pinching the $20 bill, he reaches his left hand into his left pocket and pulls out a lighter. He holds the flame up to the bill. It seems to straighten a bit, get "crispy," as he puts it, though it doesn't catch fire. He's making a gentle show of his openness. Just a lighter and a bill. Just an ordinary guy. He puts the lighter back in his pocket; he must assume all eyes are trying to watch for any sleight of hand.

He shows the warmed-up bill to the woman who gave it to him. She attests to its ordinariness.

Then he balances the bill on his fingers. Then his palm. Then on three fingers. And then he turns his hand upside down so that the bill sticks to his fingers even as they're pointed downward, seeming to defy the pull of gravity for it to drop to the ground.

"This is not magic," Patrick says. "It's science."

It's impossible not to think that he had some substance on his hands, some chemical that is causing the bill to stick.

And then he takes both hands away from the $20 bill.

And it appears to float in the air.

There is a gasp or two from the small group.

After a few seconds, Patrick blows on the bill and it floats to the ground, landing at the feet of the woman who initially handed it to him. He watches the audience look on in amazement. The woman leans down and picks up the money and shows everyone that, yes, it's just a plain old American twenty, unaltered, not sticky. There's nothing that would seem to make this bill float in the air.

Dr. Gazzaley professes to be a little conflicted by what he's seen. Like others there, he wonders whether something magic has happened and marvels at Patrick's artistry. He's also seeing the trick through another lens, a scientific one. It shows the fragility of our attention systems. Somehow, Patrick has mastered it without our awareness that he's done so. He's overtaken our top-down attention. He's got his audience thinking less and less about whether he's pulling a fast one and more about the marvelous thing they're about to see. He's transformed his goals into our goals. How will this drama end?

But, at the same time, he's perhaps toying with our bottom-up system. Somehow. By simple movements—a look here, a redirection there—he plays to the reptilian parts of our brain that can't resist novelty, maybe important stimulation. Does a movement cause us to look away from a sleight of hand at just the right second? Does he prompt a laugh that makes it hard to focus on ferreting out the mystery?

This demonstration set the stage for the next wave of attention science, and the new discoveries as the twentieth century wore on.

"We're both interested in the same thing, but we're going at it in different ways," Dr. Gazzaley tells Patrick. "We both want to understand how attention works. I'm trying to do it by looking inside the brain."

The magic trick also hints at why technology plays so powerfully to our brains, and our attention systems.

In the latter half of the century, as technology moved from the military and corporate settings into our businesses and homes, the devices grew masterful at capturing our attention. It's not that they necessarily stole it from us or pulled a fast one. But the reasons we're so swept up, researchers began to understand, is because personal communications devices are unprecedented at capturing both our top-down and bottom-up attention systems, even without our awareness.

They bring us stories like Patrick the magician, engaging the top-down systems that want to find answers, complete tasks, follow narratives. They are the narrative of our lives, our work, our relationships. How will things turn out? How are we doing at work? How are things going with our spouse, or partner, or children?

And the devices do it with lights and sounds that capture us beyond our ability to control. They buzz with incoming information, chime, change colors and images that call to us. This can reinforce our goals—alerting us to important information—but also capture our attention even when we don't want it to, even when it's dangerous, like when we're behind the wheel.

Plus, Dr. Gazzaley notes, the technology companies who build the magical gadgets have every incentive to make them our irresistible companions.

"Technology companies are trying to get more of our brains per unit time. It's as close to a business model as you can imagine. The more engaged you are in what they create, the more successful they are.

"They're driven to figure out how to engage us as immersively as possible, as deeply."

At the beginning of the twenty-first century, scientists like Dr. Gazzaley, following the footsteps laid down by 150 years of pioneers, are using technology to home in on how our attention systems work, and they are doing so, like their predecessors, in no small part because they realize technology is putting heavy pressure on those systems, challenging the ability of its human creators to keep up.

And there is another critical piece, and another key scientist. His name is Dr. David Strayer. He's a friend and collaborator of Dr. Gazzaley. He's from Utah. And he has become the most prominent researcher to apply 150 years of attention science to a new question: How does all this technology impact the ability of drivers to focus when they're behind the wheel?

CHAPTER 12

REGGIE

IN LATE OCTOBER, REGGIE took a job. It was at Murdock Chevrolet, a dealership two minutes from his house. That's where the family had purchased the Chevy Tahoe Reggie had been driving the day of the accident, though he didn't much think of that fact.

He'd gotten the position through his childhood baseball coach, Alan Williams, who worked as a salesman at the dealership. They had a position for a car detailer because one of the regular guys, Tyson, was about to leave on a mission.

Even though it was winter, Reggie wore gym shorts and a ratty, brown, hooded sweatshirt. Tyson, the outgoing detailer and a former running back and football teammate of Reggie, explained the ropes. In a small garage on the dealership property, the detailers used little blue rags to make spotless the used or new cars and ready them for sale. Brushes, screwdrivers, and spray bottles hung on small hooks along the walls.

They had an old blue power washer for the outside that was supposed to heat up the water, but it didn't. And it was unwieldy, like two hundred pounds, and so the detailers couldn't really move it. It was always humid, the steam unable to escape from the small room. Moisture damage caused the tiles on the ceiling to erode or fall off. Even in the dead of winter, Reggie tried to keep the garage door cracked to let in cool, breathable air.

But he wasn't really focused on his comfort. He was feeling deeply lost. If Tyson was there, the pair would talk, but when he left, Reggie would listen to the radio and totally zone out, sometimes drying the same spot for five minutes.

Occasionally, he'd walk across the parking lot to Feldman's & Bear River Printing. It was owned and run by an older couple. Reggie and his friends would joke that you could go there and ask for a copy and then wait forty minutes for it; in the meantime, you'd order an ice cream and get so impatient watching the woman try to dig out a scoop for ten minutes that you'd want to go behind the counter and help. These days Reggie could also be just fine with sitting there, glazed over.

His brain was lost on Valley View Road, prompting the inquiries of his state of mind.

"Are you okay?" someone would ask, maybe Alan or one of the other detailers.

"Of course."

Stay busy, he told himself. *Eventually, I'll work my way out of it.*

KEITH'S WIDOW, LEILA, RETURNED to her bookkeeping job at Tec Electric, a local electrical contractor. The company had an insurance agent named Joe and he'd heard about the accident and told Leila that she should consult with an attorney before accepting any insurance payment. Just to make sure she was getting the best possible deal and not accepting anything less than she deserved.

There was a big firm in town, relatively big, called Hillyard, Anderson & Olsen. The first principal, Lyle Hillyard, was a state senator. The third principal, Herm Olsen, was known as one of the best criminal defense lawyers going.

Leila called and got an appointment at the converted residence that the firm used for an office. Herm Olsen, tall and thin with salt-and-pepper hair, sat across his desk and started listening to Leila's story. But not far into her recounting of the accident that killed Keith, Olsen said: "I know all about it."

"You do?"

"The Shaws have already approached me to represent Reggie. Twice."

The attorney explained that, shortly after the accident, he'd been approached by Reggie's parents. But he turned down the case. The parents made a second attempt to retain him but he declined.

Leila was confused. "I'm thinking: 'This is just an accident. Why is Reggie looking for an attorney?' It was so incongruous."

But she didn't think much more about it than that. Olsen then listened patiently and told her there wasn't much to be done. A lawsuit probably wouldn't make much sense; Reggie didn't have a lot of money and there wasn't any clear finding of fault. The lawyer suggested that Leila keep him apprised of anything happening in the case, such as it was.

A FEW DAYS LATER, Leila mustered the courage to return to the scene of the accident. It was a beautiful day, midmorning. She pulled onto a side road, not far from the wreck site, that led to a gun range. Most of the accident debris had been cleared away, but signs of violence remained: blue paint from Jim's car on a bent post on the right side of the road; fragments of glass and taillights in the gully where the car came to rest; skid marks.

Leila, though somewhat in a daze, noticed something else: There was no shoulder on the road, no easy place for her to walk; no room for the cars to safely swerve. It seemed dangerous to her, even if she didn't quite put so fine a point on her observation. Just a passing thought.

She started to sob.

A man in a sedan pulled over.

"Are you okay?"

"I'm fine."

She wasn't, of course. A few days later, on October 19, she woke up vomiting. She was so weak she could barely get to the door to open it for her daughter, who took Leila to the emergency room at Cache Valley Specialty Hospital. It didn't seem like the flu or any obvious bug. It was grief. Leila was having trouble standing up. In the emergency room, they used an IV to give her five liters of saline.

A WEEK OR SO later, Jackie Furfaro, Jim's widow, took her own symbolic, difficult step. She revived Moonrise, a night elf hunter.

To do so meant going down to the basement and turning on one of the two personal computers, hers. The other of the pair belonged to Jim, and the couple often used them in tandem to play World of Warcraft together. Jim's chief character was Twinger, a gnome mage. Jackie favored Moonrise.

In the game, even as they sat at different computers, they palled around, hanging out mostly together, going on virtual quests. Jackie had now returned alone, seeking comfort in the virtual world.

Not long after Jackie logged on that night, she got a "whisper"—a message from another player. It was from Gary Maloney, an old friend of Jackie and Jim's from the Colorado School of Mines. He'd been Jim's best man. In the game that night, he was a human mage, a wizard; in the real world, he worked at a motorcycle repair shop in Indiana.

He asked how Jackie was doing.

Not great. But she wasn't telling people that. She was still grieving in her private moments, in the shower, late at night, or in the morning. She was becoming resolute that Jim, who had such a zest for life, would've wanted her to forge ahead.

That night, in World of Warcraft, her natural first reaction was to ask how Gary was doing with Jim's death. Then, for whatever reason, she let down her guard a touch.

"I'm feeling alone," she recalls writing to him. "I'm having a hard time doing it all."

Gary wrote back: "You are not alone."

A part of her, the engineer side, was still looking for some explanation; was there anything Jim could've done to avoid the wreck? Could he have swerved?

And adding to her irritation was the suggestion—the insane proposition—that somehow Jim had been at fault. That's an idea that came to her a week or so earlier when she'd gotten a visit from a local sheriff, Sergeant Tony Hudson. He was helping Trooper Rindlisbacher on the case.

Jackie had called the sheriff's office because she needed a copy of the police report from the accident to process Jim's life insurance policy. "Of course, come by and pick it up," they'd said. But to get to the office meant she'd have to drive on Valley View Drive, and Jackie, as much as she was determined to move ahead with her life, wasn't ready to drive past the place where her "first true love" had taken his last breath.

Sergeant Hudson was sympathetic and agreed to bring the accident report to Jackie's office at Utah State. As he handed the document to her, Hudson told Jackie that Reggie Shaw had "lawyered up." And, moreover, he said Reggie seemed to suggest that it might've been Jim who had been the one who crossed the yellow divider. Something was starting to smolder inside her—why couldn't Reggie just call and say he was sorry?

CHAPTER 13

HUNT FOR JUSTICE

A LOCAL POLICE DETECTIVE named Mark Robinson lived in Rindlisbacher's neighborhood. A few days after the accident, he'd gotten a call from the trooper.

"Hey, Mark, can I trouble you with a question?"

"Sure, what's up?"

"How do I subpoena cell phone records?"

"I can help you. But it's a pain. Why?"

"I'm working a fatality. I think the driver who caused it was texting."

"It can be done. It's not easy."

RINDLISBACHER'S JOB WAS SUPPOSED to be finished at this point. When cases went a little cold, or were going to take time from a trooper's day job, they got passed along to a full-time investigator. In this case, the investigator was a guy named Stan Olsen.

According to a memo that Bunderson, Reggie's attorney, wrote to himself, Olsen called him on October 25. Bunderson confirmed he was representing Reggie and that his client would not be talking to law enforcement.

The memo reads: "Olsen seemed fine with that, but he did say that the trooper is really pushing the matter."

BUNDERSON HAD ASKED THE Shaws if Reggie was texting. Reggie had told his mother he wasn't on the phone. He seemed sure of that. That's what she told Bunderson. Per Bunderson's request, she got a copy of the phone bill. It was early November.

The bill wasn't easy to parse. There were four phones attached to the account. As she read the bill, though, there didn't appear to be any records of the texts. The bill seemed only to include phone records. She took that as evidence that there hadn't been any texts.

This looks okay, she thought.

"We didn't really question whether this was all of the bill," she says in retrospect. And, she adds, there wasn't much critical questioning of Reggie on her part.

She gave the bill to Bunderson. He didn't note any problems with it. She was mollified, mostly.

"You want to trust your kids," she says, looking back.

Ed, on the other hand, couldn't stop worrying, though he didn't say it aloud. He kept waiting for another shoe to drop. He just knew it was out there.

A FEW DAYS BEFORE Thanksgiving, Rindlisbacher visited the stately Cache County Attorney's Office on Main Street. The trooper had come to see one of the seven county attorneys, a guy named Tony C. Baird. The trooper found Baird standing near the reception area, glancing at some paperwork.

"Hey, Tony. Can I chat with you for a second?"

Baird led Rindlisbacher through the frosted glass door into his office. The pair stood and chatted beside two armchairs, beneath one of the dark wooden arches that are found throughout the grand building.

Baird was around five foot seven and trim but muscular, with the body of a fitness devotee, short hair, a square jaw, and almost preternaturally straight teeth. He had an all-American face that only slightly betrayed the man's competitive nature. An opposing counsel or a defendant might be rightfully wary that Baird could be a tough foe if provoked.

Baird wore his standard fare, a black suit and white shirt he got from a department store. It was a lot fancier than his father imagined, even hoped, that Baird would wear to work.

Baird had been groomed to take over the family dairy farm, located in Lewiston, a small farming town on the border of Utah and Idaho. "Dad threatened me with taking over the family farm, so I decided I'd try some of that college stuff," explains Baird. He went to Utah State. Then his dad tried to get him to come back to the farm. "So I decided to try some of that law school stuff."

After graduating from BYU law school, he worked in several county attorney's offices in the state, spent a year as the Logan city prosecutor, and in 1997, he joined the Cache County office. By 2006, he was chief criminal deputy.

Baird, who got up regularly at 4:30 a.m. to train for triathlons, could appreciate Trooper Rindlisbacher's tenacity. "He had a bit of a reputation," Baird said. He'd heard that from Rindlisbacher himself. "He kind of goes after people.

"But I wouldn't have characterized it as overzealous."

That morning in late November, Rindlisbacher arrived with a common request: court approval to further investigate the Shaw accident. Specifically, he wanted to subpoena Reggie's phone records. He briefed Baird on the case and added that he'd seen Reggie texting in the police car on the way to the hospital. Rindlisbacher said he thought Reggie was lying.

Baird took it in. If the request for subpoena power was commonplace, the circumstances were not. Texting and driving? Sure, he'd heard a bit about it here or there, but never in the legal context.

"It wasn't something that had ever come across my desk. It wasn't something we'd ever looked at before," he says. He thought Rindlisbacher's focus on the issue was "interesting," a word that almost seems to suggest that it was a bit of a wild-goose chase. Baird wondered: *Is there even any law on the books that would allow us to pursue this case?*

His gut instinct was that this case wouldn't go anywhere. But Baird liked Rindlisbacher. And, besides, the legal standard for getting the court to allow the subpoena wasn't particularly high. It was "good cause," less than "probable cause," a standard aimed at allowing the police to pursue the next level of investigation.

Sure, Baird told Rindlisbacher. He asked the trooper to get together the affidavit facts to be submitted with the request to the court.

Rindlisbacher was going to get his phone records—and, he hoped, his proof.

CHAPTER 14

THE NEUROSCIENTISTS

A S IN SO MANY other fields, the study of attention broadened and deepened as the twentieth century sped along. And much of the gain owed to technology. Using emerging high-tech techniques, researchers pieced together the physical structures, down to the cellular level, involved with attention. Some of these developments were succinctly summarized in a book by Dr. Posner, the contemporary legend from the University of Oregon.

In the 1970s, for instance, researchers used microelectrodes to home in on a key part of the brain—called the parietal lobe—crucial to shifts in attention. In fact, as Dr. Posner would later help discover, patients with lesions to this part of the brain had more trouble shifting attention.

Use of electrodes also allowed researchers to measure the time that it takes the brain to reallocate resources when attention shifts. Say someone saw a flash of light. About one hundred milliseconds after the introduction of this new visual stimulation, the person showed changes at the neurological level. Measures like these added precision to the understanding of how long it takes the brain to react, findings that were very much in the spirit of Helmholtz and mental chronometry, the study of brain structures.

In the ensuing years, better technology meant better imaging—positron emission tomography (PET) scans, MRIs, fMRIs, EEGs. The technology allowed researchers to examine finer slices of the neurological networks, their sum and their parts. The network included parts of the brain like the anterior cingulate cortex, the dorsolateral prefrontal cortex, and, crucially, the prefrontal cortex. These areas would show changes—like increases and decreases in blood flow—when study subjects were trying to focus on something, or when they were overwhelmed with information, putting intensified pressure on the attention networks.

In Dr. Posner's book, *Attention in a Social World*, he writes that "it is now possible to view attention much more concretely as an organ system, with its own functional anatomy."

Scientists learned more about the capacities and limitations of vision, hearing, reaction time, the networks and parts of the brain most central to attention (and, in turn, to learning) and working memory—the short-term store of memory that people call upon as they maneuver through life. They broke attention down to component parts, or subnetworks, that are somewhat independent but not mutually exclusive: how you can control attention, how you sustain it, how you gather information and put it to use, how you inhibit interruption and keep out irrelevant information. And they began to understand the neural basis of the activity. At the Massachusetts Institute of Technology, researchers, working largely with monkeys, identified a powerful, multifunctional type of neuron in the prefrontal cortex. These executive control neurons, unlike many other specialized neurons in more primitive parts of the brain (a neuron in the visual cortex, for instance, might have the singular role of identifying the color red) seem to have the function of bringing together lots of information from disparate parts of the brain and organizing them, helping set direction, goal, and focus.

At Princeton, researchers used imaging techniques in monkeys and humans to validate and further the understanding of what happens when the brain is asked to consider two different sources of information. When this happens, it appears to create a state of competition inside the brain. The source of visual stimulation that is most relevant to the person gets the lion's share of the neurological resources. At the same time, as more neurological resources are put to attending to this "relevant" information, there appears to be a decline in the neurological resources put toward attending to the "less relevant" information. That might sound obvious. But the implications are significant. The research gets close to answering a very crucial question, particularly for the digital era: When a person is paying attention to one thing, do they *automatically* ignore other things, or is there a mechanism for allowing the person to modulate or control how much attention they pay to the other source of stimulation?

A hypothesis born out of the Princeton work is that attention is a finite resource: Focusing on one source (a person, a mobile device, the road ahead of you, etc.) comes at the cost of lost awareness of everything else. If that hypothesis holds, we face real challenges when we focus on a phone while driving; we can't just will ourselves to concentrate on both things because our brains are designed to put a huge emphasis on what we deem relevant information, to the point of suppressing brain activity devoted to something else.

Despite this research, Dr. Gazzaley believes this is a very open-ended question, and a crucial one. He thinks an argument can be made that the brain might be trained in its ability not just to attend but even to multitask. That's another of the key emerging areas of science: Researchers explored the underlying mechanisms of focus, they also started to look at pushing the limits of attention. In other words, can the ability to focus, once more fully understood, be expanded?

"Is there a limit to how good we can get?" Dr. Gazzaley asked one sunny afternoon in the summer of 2013, sitting in his office at UCSF. Over his desk, a row of books: *The Handbook of Aging and Cognition; Attention and Time; The Cognitive Neuroscience of Working Memory.* Beside them, a small kitschy bronze-colored statue of a monkey looking at a brain. In the corner of the desk, behind one of his two monitors, a Buddha statue.

On a whiteboard, hanging next to his desk, scribbles of blue and red marker reflect a brainstorming meeting he had earlier in the day with a postdoctoral student working on experiments aimed at identifying the neurological basis for how people distribute their attention. Dr. Gazzaley explained that people tend to be good at focusing and doing so narrowly—in a relatively small physical space—but when they seek to attend to a broader space, they lose detail.

The experiments use video game technology to challenge people to distribute attention, then use imaging to measure brain activity during the task. Could they teach people to distribute attention more widely?

Dr. Gazzaley wore a black shirt and pants and black leather boots with a silver zipper up the side. A white five o'clock shadow crossed his face, and he looked tired, but he smiled. He said he had some good news.

A study he'd been working on for four years could any day be accepted into *Nature*, perhaps the most prestigious journal in his field. The study explored whether specialized video games might be used to train older adults to do a better job at juggling two tasks at once, and might even help them become more focused over the long run. The study looked specifically at how older adults could be taught to improve attention using a driving simulator.

"Getting into *Nature* for me is like winning an Emmy," he said. He hoped to know in a few days. It would be great attention for Dr. Gazzaley's efforts, personal validation. It would also validate this emerging field of study. Indeed, even to be seriously considered by *Nature* underscored the real-world applications of this new generation of neuroscience.

The new body of work tied together the powerful new tools for looking inside the brain, with the seminal discoveries of people like Dr. Posner, with a line of science dating back to Broadbent and Treisman and their peers. Their work had focused on aviation; after all, airplane cockpits were the place where human beings were most acutely confronted with new technologies, powerful ones that taxed our neurological limitations. The cost of failure in the air was astronomical, in lives and dollars.

From this rich tradition of aviation attention science—dating back to World War II—there grew a new field. It was the application of attention science to driving. One scientist in particular, David Strayer, played a pioneering role, though he faced great opposition when he first started his work.

IN 1989, DR. STRAYER earned a PhD in psychology from the University of Illinois at Urbana-Champaign. He was developing a specialty in how people become experts and how they acquire skills, and how those skills get compromised. How do they process information? How does information overwhelm them?

The University of Illinois was one of the foremost places in the world to look at the interactions between humans and technology, an area of study known as "human factors." The university had a reputation for spawning world-class scientists exploring how to optimize use of technology in airplanes and in the military, in the tradition of Broadbent and Dr. Treisman. The basic idea: How to make machines work best for people without becoming overwhelming.

That's what Dr. Strayer figured he would do. In 1990, he went to work for GTE Laboratories. GTE was a telecommunications giant that did a lot of work in the consumer market, as well as for the government, including the military.

And this was wartime. In January 1991, the United States entered the first Gulf War and was going after Saddam Hussein. As in prior wars, this one presented scientists with the challenge of helping soldiers and their commanders figure out how to take advantage of technology rather than be overwhelmed by it. But the technology issues were no longer just in the cockpit, affecting only pilots, they were everywhere.

Communications tools and networks had become essential to success. Guys like Dr. Strayer were hired to figure out how to configure networks and displays so that the information was useful and life-preserving—not overwhelming and deadly.

Dr. Strayer quickly realized that this work had widespread applications. And so, while at GTE, he started noodling a nonmilitary question, one that troubled him and he couldn't shake. It had to do with the consumer side of GTE's business, the part that was making mobile phones and had begun marketing them for cars. (Eventually, GTE would be acquired by Verizon, one of the biggest mobile phone providers in the world.) To Dr. Strayer, the idea of using a phone in a car was troubling, at least based on now decades of science establishing the limitations faced by pilots when they taxed their brains with too many visual inputs, sounds, and physical demands.

He went to a supervisor at GTE and said, "Everything we know from aviation psychology indicates this is likely to be troublesome," referring to the idea of a car phone. "Before we start marketing this, we should think about it."

Shortly thereafter, he heard from his supervisor that the company leadership wasn't particularly interested in pursuing the safety question.

"Why would we want to know this?" Dr. Strayer was told. "That will not help us sell anything."

Drivers were, arguably, the most important early market for mobile telecommunications. In fact, the first big commercial push for mobile phones were car phones.

According to an article published in the *New York Times*, mobile phone companies, going back as far as the early 1980s, openly marketed the devices for use by drivers. One ad in 1984 asked, "Can your secretary take dictation at 55 mph?" The mobile companies concentrated their cell phone sites along highways, hoping to capture business from bored motorists. They succeeded. A longtime telecommunications analyst named Kevin Roe told the paper that 75 percent or more of wireless company revenue came from drivers well into the 1990s. "That was the business," he was quoted as saying in the *Times*. Wireless companies "designed everything to keep people talking in their cars."

Seeing these trends in the early 1990s, Dr. Strayer decided he was interested in getting answers—with or without his company's participation.

He needed to be in a place where he could pursue answers. That turned out to be the University of Utah, in Salt Lake City. He got an assistant professorship and started putting together experimental protocols, trying to borrow from the masters who'd come before him—Broadbent, Treisman, and on and on—essentially adapting the experiments that had been done for pilots to car drivers and car phones.

There was nothing revelatory about such an application. It made sense. But on a societal level, the research was nothing short of profound. After all, scientists had been focused for so long on technology used by an elite class of people—not just pilots, but the soldiers and others who were privileged to have access to the supernatural devices.

But now the devices were becoming part of everyday life. Or they were poised to become so. Cockpits in cars, Dr. Strayer thought.

Perhaps it was not surprising that Dr. Strayer couldn't find a lot of funding for his research, given how many financial interests were arrayed in the other direction. So he did his first experiment at the University of Utah by cobbling together about $30 in parts. He built a primitive driving simulator.

In the experiment, the subject (an undergrad from the university) would sit in a chair holding a joystick, which controlled a car on a computer screen. The subject was told to do simple tasks: (1) follow along a curvy road, and (2) hit a button if a red light came on. That would cause the car to brake. At the same time, Dr. Strayer asked the subjects to talk on a cell phone. Some used a handheld phone, some hands-free, with a headset. Separately, he also had the subjects "drive" while listening to the radio or a book on tape.

There was a huge difference in the "drivers'" results, depending on what activity they were doing. When they were talking on a phone they made twice the number of errors as when they were listening to the radio. The error rate was higher both when using a handheld and hands-free phone.

Dr. Strayer says he was struck by the results: "There's something about the phone conversation that's really kind of unique."

In 2001, he presented his findings at an annual meeting of the Psychonomic Society; it was the very first study to show the effects of talking on a cell phone. The findings were very well received. Dr. Strayer had made an important connection between the past research in the attention/distraction field and the challenges faced by multitasking drivers.

"We made a link between this and fifty years of research on attention and aviation," he says. "The attentional limits we saw with pilots apply to drivers of cars. It was an important first step."

At the time, though, Dr. Strayer thought it might be a last step, too—a final statement. He didn't realize the pace at which mobile phones would be adopted, and the extent to which they'd be used in all walks of life, all the time. Plus, he said he couldn't find anyone to adequately fund the research; there was so much cultural momentum driving adoption and so many powerful business interests that no one really wanted to find out—or was interested in paying to find out—that the magical new technology had some side effects.

And he was just beginning to understand something that would become much clearer as his research grew: The devices took a driver's *mind* off the road, even when hands were on the wheel and eyes were looking ahead.

"If you asked at the time, I'd have said it's mainly a visual and manual demand. But it's a visual, manual, cognitive problem."

IN LOTS OF SCIENCE fields, the researchers get to know one another's work and often collaborate. That has happened in recent years around the field of attention, particularly the study of attention and its relationship to technology. There were more scientists going into related fields, and more money. A new subspecialty was emerging as scientists grappled with the onslaught of new devices. Within this new ecosystem, Dr. Gazzaley and Dr. Strayer crossed paths.

The two had been invited to participate in a small one-day meeting at the campus of Stanford University in Palo Alto. The host was Clifford Nass, a sociologist who had been a math prodigy as a child in New Jersey and whose plan was to become a computer scientist. But Nass's life was changed by a fatal accident that took place in Utah, when his brother, traveling across country, was killed by a drunk driver. The event devastated Nass and his family and ultimately had him reevaluating his life choices. He wound up studying sociology at Princeton and then becoming a professor at Stanford.

In February 2011, he invited Strayer and Gazzaley to join him, along with Anthony Wagner, a Stanford psychologist; Gary Small, a psychiatrist at UCLA; and Daphne Bavelier, a cognitive scientist at the University of Rochester. Each of them was gaining notoriety for understanding how heavy technology use—multitasking—impacted the brain. Nass wrote to the group: "My goal in inviting you is simply to bring together the people who are doing the most exciting research in multitasking."

Dr. Strayer and Dr. Gazzaley hit it off, even though they seem cut from entirely different cloth, both figuratively, with respect to personality, and literally, in terms of their sartorial choices. Dr. Gazzaley favors sleek shades of black, and Dr. Strayer is more of a Levi's and T-shirt kind of guy. They decided to try to collaborate. Dr. Strayer would bring to the table his research into the behavior of drivers. Dr. Gazzaley, a neurologist, would bring his understanding of neural networks, and the technology that looked at the inner-workings of the brain.

"His techniques are bleeding edge," Dr. Strayer says of Dr. Gazzaley. "We were able to take our leading-edge research in driver distraction and pair it with cutting-edge neuroscience."

Dr. Strayer wanted the emerging field of research to also answer another question: Why, given it was becoming clear that the brain faced limitations, were people continuing to multitask, particularly in challenging, even dangerous, situations? When he first started his work on distracted driving, he just assumed people would stop the behavior when they realized how dangerous it could be. But when phone use by drivers continued, even grew, he was forced to reach another conclusion, one that vexed him. People didn't stop using the technology, because they couldn't.

"I assumed people would come to their senses," he says. "It was naive on my part. Still to this day, I'm surprised by how addictive and how alluring the technology is."

The stage was set to incorporate the decades of past research, and to marry behavioral science with neuroimaging, to answer a new question: Why does interactive media do such an extraordinary job of capturing our attention?

CHAPTER 15

TERRYL

LATE IN THE FALL, Jackie Furfaro drove her daughter Stephanie to gymnastics practice at Air-Bound, the gym on Main Street in Logan. It was just weeks after the accident, but Stephanie insisted on continuing to practice. Jackie thought it best to try to live life, persevere, not look back.

Not that she wasn't bewildered and furious. She just continued to express her grief in private. Sobbing while no one else looked.

But someone *was* looking, quietly paying attention.

After gymnastics practice that day, she approached Jackie in the parking lot. Jackie knew the bubbly blond woman whose daughter went to gymnastics with Stephanie.

"Hi, Terryl," said Jackie.

"Jackie. I'm so sorry."

Terryl had come a long way. She was Terryl Warner now, married, with kids, and a track record for taking on tough cases as a victim's advocate. In this casual exchange with Jackie, she was on the cusp of taking on a case that would test her willingness to push through obstacles, setbacks, authority figures.

AFTER SHE GRADUATED FROM USC, Terryl was in an LDS church group in Irvine when she met April, a gorgeous, charismatic woman, tall and slender with long brunette hair. The pair started chatting. Within days, even within hours, something clicked. April was a recent graduate of BYU who modeled for a local diet center where she worked. Terryl felt very comfortable with April, so much so that Terryl felt she had a best friend for the first time in her life.

But something was bothering April, too—she was tired, had dark circles under her eyes. In September, April called Terryl.

"Something's wrong with my blood."

Within days, April was admitted to the hospital with leukemia. Terryl dropped out of the Bank of America management training program so she could spend time at the hospital. She found temp work at legal agencies, keeping alive that distant dream of going to law school. She spent many hours at the hospital, St. Joseph's, in the cancer ward, climbing into bed with April, talking about life and its purpose, and religion. April had dreamed of going on an LDS mission. Terryl had never really thought about it, but was starting to.

After her diagnosis, April had gotten engaged to Neal Harris, a BYU grad who worked hard and was intensely devoted to his friends. Terryl watched how April was treated by Neal—the way he climbed into bed with his bald, bone-thin fiancée—and rubbed the parts of her body that ached from chemo. Neal brushed April's teeth with the little spongy pink toothbrush that was too soft to cut into her fragile gums.

It made a big impression on Terryl. "He taught me: It's okay. Not every guy is bad."

The respect was mutual. To Neal, Terryl seemed so full of life, unfazed, really, even fearless. She spoke her mind. She wasn't put off by April's beauty or presence, the kind of thing that made some women uncomfortable.

"She wasn't intimidated by anything," says Neal.

TERRYL, AFTER MUCH CONSULTATION with April, decided to go on an LDS mission, to Costa Rica, in July 1989. Her eyes were wide open with awareness. She knew she was helping fulfill April's dream. And Terryl felt she was taking a step toward making her own destiny. Yet she also knew she was escaping. She wouldn't be there when April died. But that seemed to be okay with April, who implored Terryl to go.

Terryl also knew she'd likely miss the deaths of her maternal grandparents, who'd been among the handful of people to look out for her when things were really at their worst. When she went on the mission, she recalls, "it was a realization for me that God loved me and wanted something good for me and it was okay to have a good life and to want to help others." But "it was also something of a protector for me. I had this best friend, and she was going to die. And these other people in my life I was close with, and they were going to die."

All three passed away while Terryl was gone.

She came home from her mission on January 2, 1991. She had enough distance now from her upbringing that she felt it wouldn't define her for the rest of her life. Maybe she could even use her upbringing to her advantage.

THE WOMAN LOOKED TO be in her thirties, with dark hair and dark skin. A feral, terrified look in her dark eyes, wet with tears. No wonder. She had holes burned in her face, like someone had taken a cigarette to her. Except much worse, if it's possible to imagine: a strange pattern of dark, round burn marks all along her cheek.

Terryl sat at her desk in the small offices of the community services victim's advocate program located in the Orange County courthouse. It was 1992, and this victim didn't want to be here. Behind sobs, she, and the friend who dragged her in to see Terryl, explained what had happened.

The woman had four children all under the age of twelve, and a jealous, controlling husband. He didn't like her leaving the house, certainly not using the car. He didn't even like her driving to her job at Taco Bell, even on a rainy night. So sometimes, when it rained, the woman would ask for a ride home from a colleague.

That's what had happened two nights earlier. She'd finished her Taco Bell shift. It was pouring. The manager offered to give her a ride home when his shift ended, a half hour later. In a downpour, he dropped her off in front of her house. Her husband, angry she was late, went into a rage upon seeing another man drop her off.

He ripped her hair, he tied her up. He said: "I'll make it so no man will ever want to look at you again."

He held the clothes iron to the side of her face.

She didn't call the police. She said she didn't want to be there, talking to Terryl, dealing with it, confronting her husband. Terryl recalls the woman imploring: "I'm a religious woman, and my husband, he's an important man." She explained: He's the bishop of the local Mormon church, an LDS lay leader.

"Nobody has a right to hurt you. Nobody," Terryl says she told the woman. "You deserve to live a normal life."

Inside, Terryl seethed. She tried to keep her wits about her. The woman wouldn't relent. Finally, Terryl came to the rationale she felt inside so personally, so deeply, the reason that, if nothing else, was her best bet at swaying women to act.

"If you don't do something, eventually your kids will be the punching bags."

The woman agreed to go ahead. They served a protective order so the man couldn't go in the house, and a restraining order to keep him a safe distance away from her. Eventually, she got a divorce. It was one of the victories, weighed against plenty of losses. To Neal Harris, now one of Terryl's close friends, it was a powerful example of the leather skin underneath her bubbly exterior.

"It was one of the most awful things I'd ever heard about," he says, looking back. Not just the act, the actor. "He was a Mormon bishop, an ecclesiastical leader in her faith. It didn't matter," he says, not that it should have. "She comes in with the attitude that we are going to cut his nuts off—basically, this isn't a time for crying, this is a time to make sure he's punished for what he did."

ONE NIGHT A FEW years earlier, in Norwalk—a tough Los Angeles– area city checkered with Locos and Primos and five or six other rival, mostly Latino, gangs—a high school party broke out, and a fight along with it. Words were exchanged, egos bruised, reputations challenged. Seventeen-year-old Alan Warner threw a big roundhouse. He crumpled the other boy, breaking his jaw.

Alan had had his share of scuffles and could handle himself okay. He felt he had to, given the neighborhood he grew up in. But it wasn't just that. He wasn't much into toeing the line. He was drunk the night of that fight, and plenty of other nights, too. He tried pretty much every drug.

He never looked twice at a girl like Terryl.

"She was straitlaced and I was on the other end of the spectrum," he says.

He knew her through the Cerritos church. And he knew her family a bit. She was short and kind of attractive.

"I was looking for someone who could keep up on the weekends."

BEFORE TERRYL'S MISSION, HER mother suggested she go on a date with the same Alan Warner. "Remember him?" her mom had said. "He used to help us move furniture when we moved. Dad is a roofer, one of six kids, nice Mormon family." Yeah, Terryl knew him. He was rough around the edges, a football player, the guy who'd made fun of her car in the church parking lot after she'd gotten in a fender bender.

But he was going on a mission, too, his to Alaska. He'd gone through a repentance process so he could cleanse himself of the behavior that would preclude him from going on a mission. While on it, he and Terryl wrote a few letters to each other.

When he came home, he asked Terryl on a date. She invited him over for a home-cooked dinner. He loved it. Only later did he learn the cooking was actually done by Terryl's roommate. But he was hooked, not on the food so much, but on this strong-willed, determined young woman. She could push him in the right direction, he thought. Plus, he wanted a family.

So did Terryl. She'd reached a kind of peace about it. She could will it. She could make everything right.

Two months later, on Labor Day 1992, they were engaged. And seven and a half months later, they got married at Los Angeles's Mormon temple, like Terryl had dreamed about as a little girl. At the wedding reception, held at the Los Coyotes Country Club, Danny showed up belligerent, but a handful of men were at the door, waiting for him, and turned him away.

As determined as Terryl had become to not be weighed down by her childhood, Danny remained a daunting and unpredictable figure; and her brother was sliding, disappearing into the maw of addiction; bad luck seemed to follow her, the death of April included.

There had to be a way out, for good.

One day, not long after Terryl and Alan were married, he came home hurting after one of those long, rough days roofing. He'd thrown up over the side of one of the houses, sick.

"She said: 'You're not going to do this anymore. You're going to go to college and we're going to get you into another profession.'"

They found the right college: Utah State in Logan. Alan had family in the area. Both were Mormon. It was beautiful country. Alan thought maybe he'd become an electrical engineer.

Before moving, Terryl faced her biggest challenge yet, the one that most terrified her. She and Alan had a baby, a little girl, Jayme, born on September 15, 1994. That meant Terryl, the little girl from the broken home, would be the mom trying to keep the pattern from reproducing itself. She had a new husband who sometimes needed pushing; Danny was still hovering out there; and Michael, his own drug use intensifying, was becoming a bigger challenge in her life, a wild card. How could she break the cycle?

SHE FELT A LITTLE more solid on the work front. Within a few months of moving to Logan, Terryl found a job. She became the victim's advocate with Cache County, in the prosecutor's office. Several years into the job, a story worked its way into the prosecutor's office about police exploring allegations that a dad from a nearby small town had raped his teen daughter. The evidence wasn't clear in part because the mother was telling the daughter to let the thing drop, according to the county attorney at the time, Scott Wyatt.

Wyatt had hired Terryl and had watched her become a tenacious victim's advocate. She took on a role, Wyatt said, that wasn't typical of victim's advocates. Terryl didn't just offer advice to the victims going through the system; she got involved in the case and pushed investigators and prosecutors. Wyatt says she was right so often that he and others listened carefully to her.

"She was never overzealous, but sometimes it was like: 'Oh my gosh, can you give me a break? We've got a big job to do, can't we just move on?'"

In the case of the rape allegation, Wyatt says, Terryl began looking into it, almost taking on an investigator's role. Wyatt recalls she was "convinced there are much bigger problems" with the family. The dad got arrested, and eventually convicted. "The case would've been dropped without Terryl," Wyatt says, adding: "Terryl was the only person who supported this girl." And there was more: Once the man was imprisoned, it came out that he'd raped his other children, too.

"It happened again and again," Wyatt says of Terryl's involvement in cases. In another instance, a mother reported that her daughter had been sexually abused by a man in the community. The police couldn't nail it down. The prosecutor's office, following Terryl's instinct, used wiretap evidence to secure a conviction.

Wyatt said Terryl also helped organize community meetings with religious leaders, the bishops, who often were the first people contacted when a problem cropped up. Terryl was trying to take the stigma out of family abuse, and show it's not, among other things, something a family can just take care of itself. Wyatt says of her: "She thinks: Why spend all of our time fixing problems when we can perhaps devote a little time to preventing them in the first place?"

Here's how Terryl described her hopes and passions: "I'm not going to be like Erin Brockovich, someone who is going to do amazing things. But I'm going to do something. I'm going to do something even if I'm told no."

TERRYL HAD, OF COURSE, read and heard about what happened to Keith O'Dell and Jim Furfaro. She wasn't sure there was anything she could do professionally, because at the time she assumed the accident took place in Box Elder County, where Tremonton was located. In fact, however, the accident had taken place across the Cache County line, within Terryl's area of jurisdiction.

Regardless, she thought she might extend herself to a fellow victim, a friend. On that night in the late fall, after their daughter's gymnastics practice, she approached Jackie and made an overture that set in motion a cascade.

"I'm sorry," Terryl said to Jackie. "Is there anything I can do to help?"

PART TWO

RECKONING

CHAPTER 16

THE NEUROSCIENTISTS

OUTSIDE LAWRENCE, KANSAS, TWO dogs and a man walk along a gravel road surrounded by open fields. The sky is vast, unblocked by buildings, even the modest ranch-style homes are separated by many acres. Elm, ash, oak, and other trees dot the landscape in clumps. The two dogs are Lupin, a Boston terrier, and BeBop, a tricolor Welsh corgi. The man with their leads stuffed in the pockets of his orange fleece is Dr. Atchley, the Kansas University psychologist, an introspective former army captain turned scholar.

He's thinking about the Bible.

Earlier in the day, an unusually temperate February morning, one of Dr. Atchley's cognitive neuroscientist students had asked him about the scientific validity of intelligent design. That's the idea that the existence of the earth and its people are better explained by divine inspiration than by evolution. Dr. Atchley's initial response was that intelligent design qualifies as "pseudoscience." But he brought himself here, to this walk, in the open and the silence, to clear his head and ponder the question.

This is a big part of every day for Dr. Atchley, if not the walk, the retreat. He lives here, on twenty-six acres, about a ten-minute drive from campus—in an underground house that he designed a few years ago. It's around 2,600 square feet, and covered fully by hardscrabble brown dirt on one side, with nearly eight feet of land above the structure of the roof. One side, facing south, is window-lined, inviting light and heat.

His wife, Ruthann, is chair of the psychology department at Kansas. The Drs. Atchley think of this place as their quiet, protective cocoon, and also part architectural novelty that they still seem bemused the bank ever let them borrow to build.

Underground, inside the house, it's cool in the winter, and mostly in the summer, too. Year-round, there is no cell phone service, at least not inside. They do get Internet access. Outside: owls, frogs, deer, the occasional bobcat, and Dr. Atchley keeps a small colony of bees.

In the garage, he parks a Subaru with a personalized license plate. It reads: ATTEND. When asked to explain, he jokes: "Because 'Turn off your fucking cell phone' is too long for a license plate."

ON THE DAY OF the conversation about intelligent design, Dr. Atchley walked and thought, and later that night, he crafted an email back to his student in which he quoted Hebrews 11:1.

"Now faith is the substance of things hoped for, the evidence of things not seen."

And then Dr. Atchley continued his note in his own words: "If you try to play the game of using a method that relies on testing observable evidence, it seems to me you ignore the message of faith, which is that faith does not require evidence and should be strong in the face of evidence to the contrary (reread Job for a better lesson on faith despite contrary evidence.)"

In other words: Don't expect science to prove your faith.

Religion doesn't come up much for Dr. Atchley, and when it does, he can call on the years he spent at a Jesuit high school. Usually, he's exploring people's dedication to a different idol, that of technology.

Why are we so drawn to our devices?

What makes us check them all the time? When sitting at dinner? When behind the wheel of a car?

"Are these devices so attractive that, despite our best intentions, we cannot help ourselves?"

He believes that to be true, but is not relying on his gut. He wants to prove it. "Some of the questions appear to be difficult, even impossible to answer. But they might not be impossible to answer. We may not be able to point to the exact mechanism in the brain, but we can infer it with the right kinds of experiments."

For Dr. Atchley, who's doing some of the foremost research in the field, the questions suggest he's awoken from his own blind faith in technology.

BEFORE THERE WERE COMPUTERS in Silicon Valley, before the land gave way to industrial design, there was fruit. Oranges, pomegranates, and avocados. Acre after acre. Trees and dust under a hot sun. Perfect farming conditions. Before Silicon Valley was Silicon Valley, it was one big farm, the Valley of the Heart's Delight.

Talk about innovation. The fruit cup was invented in the Valley of the Heart's Delight, by Del Monte.

Then World War II came along. And the Varian brothers and Hewlett and Packard arrived. They got their early funding from the federal government. Defense contractors with a high-tech bent.

This intersection of computers, telecommunications, and the military would yield a change arguably as significant and characteristic of modern life as anything in medicine and the industrialization of food. It was the birth of the Internet, the product of a research program initiated in 1973 by a branch of the military called the Defense Advanced Research Projects Agency (DARPA). The aim was to create a communications system that would go across multiple networks, making it less vulnerable to attack or instability. It was, if nothing else, very hearty.

Silicon Valley became an engine for its growth, serving it and feeding from it. Still, there were orchards left, a dwindling handful. Computers and communications technology commingling with open spaces.

A perfect place to be a thirteen-year-old. Particularly one with a bike, an innate curiosity, and a latchkey.

A younger Dr. Atchley, an only child, with working parents, had to entertain himself. His mother, a hippie by orientation, worked as a legal secretary; his stepfather was a physicist and engineer who designed machines that made silicon wafers, which computer microprocessors are built on.

Left to himself, Paul, a slight boy with dark hair, spent hours trucking through the patchwork of fields and residences. He looked for rocks to turn over and ditches to explore. "I can still smell what it smelled like in those summertime dried-up algae-frog-filled catching places," he says.

He was also fascinated by computers. He vividly remembers the day he went with his dad (actually his stepfather who had legally adopted him) to the local Byte Shop to look at the first Apple computer. It cost more than $1,000. Way beyond the pale. Paul's own at-home technology was relegated to a black-and-white television he had in his room, and on which he watched local channels and monster movies. He loved fantasy and sci-fi, and read books about what to do to survive a nuclear attack; he imagined he'd live in an underground house.

Finally, he got a computer. It was called a TI-99, made by Texas Instruments, and it was one of the first home computers. Kids used them to play games. Paul did that. But his interest went beyond games.

"It wasn't that it gave me the ability to play games. It was that it gave me the ability to do anything I wanted with it—program my own games, use the tape drive to store secret information. What it really represented was limitless potential.

"At that point in my life, I was convinced I'd become a botanist on a space station."

He was serious. Technology had been moving so quickly, doubling, tripling, quadrupling in power. "We were expanding beyond ourselves, beyond our own planet, pushing back frontiers, bettering ourselves."

And communicating across geographic barriers. From his room, he could reach out across the city, the country, the globe.

He didn't realize it, but he was in the midst of an extraordinary time.

ONE REASON THINGS WERE changing so much owed to a familiar technology maxim called Moore's law. It essentially says computing power doubles every eighteen months to two years.

But there is another key technology axiom. It defines a different kind of change to the world and our lives, and indirectly would drive Paul's eventual research: Metcalfe's law. It defines the value of a telecommunications network, say, the Internet, as proportional to the square of the number of users. The more people, the more valuable the network.

It was named after Robert Metcalfe, an electrical engineer and innovator who helped develop the Ethernet computing standard used to connect computers over short distances. According to a history published by Princeton University, Metcalfe's law was officially christened in 1993, but the principle was first identified in 1980, when Paul was getting his TI-99 and dividing time between it and his bike. It's not that the concepts eloquently captured by Metcalfe's law were entirely new; networks had been developing, and their potential significance was defined in the latter half of the century. But Metcalfe put a fine point on what had become a core attribute of media by the end of the twentieth century.

One simple way to think about how much had changed, how much power had come to personal communications, is illustrated by a simple comparison. In World War II, an extraordinary calculating machine commissioned by the U.S. military, the Electronic Numerical Integrator and Computer, could, each second, perform around 350 multiplications or 5,000 simple additions. By 2012, the iPhone 4 made by Apple could do two billion instructions per second. The iPhone 5, in 2013, even more.

The ENIAC weighed thirty tons. The iPhone 5 is less than four ounces. It carries voice communications and the Internet, a crystallization of all the wondrous powers of the previous millennia, a machine in our pockets that, on its face, worked fully in people's service, the ultimate entertainment and productivity machine.

By the standard of the iPhone, when Dr. Atchley was just teenage Paul, the pace of their communications and the amount and variety of information—print, voice, video—was relatively limited (maybe not by phone but certainly by computer or mobile phone).

As Paul was growing up, a half generation before Reggie came of age, there was a coming together of these two fundamental computing principles, Moore's law and Metcalfe's law. One defined the acceleration of computer processing power, which allowed not just speed, but so many capabilities, including, at its core, interactivity; the other captured the rapid expansion of the communications network and its value.

In union, they were combining to provide unprecedented service to humans. But they were also putting a new kind of pressure on the human brain: Moore bringing increased information, ever faster, and Metcalfe making the information so personal as to make the gadgets extraordinarily seductive, even addictive.

A FEW MONTHS BEFORE Dr. Atchley took that February-morning walk with his dogs, he attended a first-of-its-kind conference in Southern California. About two hundred neuroscientists gathered with support from the National Academy of Sciences to confront a new question: What is technology doing to our brains?

The introductory lecture was given by Clifford Nass, the provocative Stanford University sociologist who two years earlier had gotten Dr. Strayer and Dr. Gazzaley together to think about the science of multitasking. Now, Dr. Nass was pushing scientists to go beyond the existing science and ask hard questions about whether the ubiquity of constantly connected mobile devices could, ultimately, hamper the things that make us most human: empathy, conflict resolution, deep thinking, and, in a way, progress itself.

Near the front sat Dr. Gazzaley, mentally preparing for his own talk to be given shortly. Seated in the upper right, Dr. Strayer wore glasses, his neck slightly hunched.

And in the back row, Dr. Atchley. His wire-rimmed glasses were perched on his nose and his Macintosh laptop was shut beneath him. That was somewhat noteworthy; many in the crowd had laptops open, including a guy just in front of Dr. Atchley who had four windows open, checking email, the news, and a shopping site.

In explaining why he doesn't open his laptop, Dr. Atchley calls upon a phrase from his Jesuit training. "Lead me not into temptation but deliver me from evil," he says, paraphrasing Matthew 6:13.

He thinks that if he opens his laptop, he'll start checking things. Get distracted from the lecture, and his own analysis of it. He doesn't trust himself to be disciplined, and he says some fundamental neuroscience has emerged to support his fears. There's plenty of anecdotal evidence, too.

At the University of Kansas, the journalism school does a periodic "media fast," in which students aren't allowed to use their devices for twenty-four hours. When the fast took place in the fall of 2011, students reflected afterward about their experience.

To wit:

"How could I abandon my closest friend, my iPhone?"

"My media fast lasted fifteen minutes before I forgot that I was fasting and checked my phone."

"The withdrawals were too much for me to handle."

"Five minutes without checking a text message is like the end of the world."

"I don't want to do this assignment again."

Why? Why is this stuff so compelling?

Dr. Atchley says one thing that makes the question fascinating to him is that when people multitask, they often do so in situations that defy common sense, say, trying to concentrate on an in-person conversation while checking a sports score or attempting to drive while dialing a phone. He believes that there are impulses driving these multitaskers that don't meet the eye. In fact, he says that increasingly technology is appealing to and preying upon deep primitive instincts, parts of us that existed aeons before the phone.

For one: the power of social connection, the need to stay in touch with friends, family, and business connections. Simple, irresistible. "It's a brain hijack machine," he says. He's trying to prove it.

This is what comes next in the study of the science of attention, the latest wave. Are there some things that can so overtake our attention systems as to be addicting? Is one of those things personal communications technology?

A trip to his lab, he says, will help illustrate his quest.

CHAPTER 17

TERRYL

TOWARD THE END OF 2006, when Terryl first approached Jackie outside of gymnastics and asked how she might help, Jackie showed typical stoicism and said: "I think we're good."

Terryl, respecting Jackie's strength and privacy, let it go. Besides, she still thought the accident had happened in the adjoining county, Box Elder. So she settled in for being a friend.

Just before Christmas that year, the Furfaros and the Warners went to a gymnastics meet in Park City. Jackie piled the girls into the Saturn. Terryl and Alan drove their van, carrying the brood of four: Jayme, the oldest, then twelve years old; Taylor, the boy, age ten at the time; Allyssa, just shy of five years old; and Katie, who was three and suffered from cystic fibrosis and autism.

At the meet, and over meals, talk between the Warners and Furfaros was about the kids and gymnastics, not the accident.

A FEW DAYS LATER, Jackie took Stephanie and Cassidy to her mother's house in Nevada for the Christmas holiday. That meant driving on Valley View Drive.

As they left home, she tried to avoid the thought that within ten minutes she'd be at the spot where Jim had died. She put on the radio and, for the girls, started a Disney movie, which showed on monitors that attached to the back of the front seats, absorbing Stephanie and Cassidy. The aftermarket DVD player had initially been a sore spot between her and Jim.

"Why do they need this?" Jim had asked.

She told him sternly: "You're not always with us on the drive; you don't know how hard it is to go six hours without entertainment."

Altogether, they were at least a nine-screen household at the time of Jim's death. There were the computers downstairs and a third on the dining room table that the girls sometimes used. The movie screens in the backseat of her Saturn. Each parent had phones. There were two televisions. Not included are the various GPS and other devices Jim played around with.

Soon after the accident, Stephanie, the older of the girls, had begun to play World of Warcraft, using her father's account and his computer and desk, peering into his hefty NEC monitor. Jim and Jackie had decided that she could use World of Warcraft when she turned six.

After Jim's death, Jackie and the girls also got lost in films, holding movie nights. In particular, they'd gather around and watch *The Sound of Music*, or another classic, and eat a take-out pizza. For a while, Stephanie suspended her playing of Dance Dance Revolution because it reminded her of her father. A few days after Jim died, at the viewing of body, Jackie had asked the girls to look at their father so they could know and understand that this was a real thing. As Stephanie would later recall in a school essay, her mother came up to her afterward and said: "I am so sorry this happened, and I know you can be strong and get through it."

Media seemed to help.

As Jackie drove that morning on their Christmas journey, the girls watching Disney, she tried to lose herself in the radio. She thought: *If I get teary, my vision gets bleary and then it'll be hard to drive and then I'll have to explain to the kids why I'm stopping.*

For Christmas that year, she felt she overdid it a bit with the presents. "I got the girls too many books."

FOR CHRISTMAS, JACKIE WAS in Nevada, Leila was at home, Terryl was in Mexico—at an orphanage.

On that Sunday before Christmas, Terryl and Alan surprised the family with the trip to Mexico. But it wasn't exactly a vacation, Terryl told the kids. She brought out a bunch of seemingly random supplies: toothbrushes, combs, little deodorant sticks. They were the makings of "hygiene" kits to be handed out, along with puzzles, books, and sun hats, at an orphanage in Puerta Peñasco.

The next day, they piled back into the Ford Windstar van and drove south, stopping for the night in Searchlight, Nevada.

Puerta Peñasco is about a hundred miles south of the Arizona border on a strip of land that connects the Baja Peninsula to the rest of Mexico. The Warners stayed free in a condo that belonged to Neal Harris, Terryl's old friend from Southern California who had been engaged to April before she died of cancer.

Neal had hit it big in the technology field, really big. He held top sales jobs at four different technology companies that were ultimately sold or went public for more than a billion dollars each. There was Syn-Optics Communications, which was among the pioneers in creating technology to allow for faster, more efficient delivery of data over Ethernet lines. It merged in 1994 with another company in a 2.4-billion-dollar deal, a precursor to the dot-com boom.

Neal moved to Ascend Communications, which built little boxes to terminate Internet signals and was acquired by Lucent Technologies in 1999 for $24 billion, one of the largest acquisitions in history. And, after that, Neal went to Foundry Networks, which made routers and switches— essential pieces of technology to deliver Internet traffic. It would be acquired by Brocade for $3 billion in July 2008.

He was, like so many fortunate Americans, a big beneficiary of

Metcalfe's law. More connections, faster connections, more efficient connections—Neal and others in the booming tech industry serving a seemingly insatiable drive to communicate with one another and trade at ever-increasing rates. They were creating the most powerful robots and conduits the world had ever seen, things that each year were making the tools of the previous year seem slow by comparison. They were creating wealth. This was another side of the technology revolution, big, big money. Neal had amassed a multimillion-dollar fortune.

Among the benefits of his wealth, he purchased two three-bedroom condominiums in Puerta Peñasco. Terryl said she wanted to take the children there for Christmas, to work at the orphanage, to teach them to give back.

"My kids need to see there are people who have less than them," she says. "They need to learn empathy at a young age."

On the outside, the Warners seemed like they'd developed into a perfect family. The children were excelling in school. They attended church. Neal looked at Terryl with some wonder, given her childhood. How could she turn out so different, given her upbringing? Would it come back to haunt her? "It's a question I end up asking myself quite often," he says. "Think about all the things she had going against her."

It helped that the Warners had moved to Utah. But not completely. There were parts of Terryl's journey the decade prior to the crash that showed how haunted she remained by her youth.

IN THE FALL OF 1998, eight years before Reggie hit the rocket scientists, Terryl stood in a Mervyns at a mall in Orem, Utah, about forty-five miles south of Salt Lake City, not far from Provo. She was glancing at the women's clothes, passing time. She and Alan had driven down to visit his family. They'd, of course, brought with them Jayme, their firstborn, and Taylor, the newest addition to their family, who had been born two years earlier, on October 18, 1996.

As Terryl was glancing at the dresses in Mervyns, she suddenly realized that Taylor had disappeared.

"Taylor?"

No response.

She was searching frantically.

"Taylor!"

No response.

"Alan! Alan!"

Her boy was missing. It felt just like Danny had snatched Mitchell. Just like in her nightmares. Was Danny there?

Alan ran over. "What's going on?"

Terryl was in a full-blown panic. And then Taylor appeared. Her little guy peeked out from between two dresses in a circular rack. She scooped him up. She was beyond crying, nearly hysterical, which was not at all Terryl's style.

"We're going home. We're going home."

They never made it to the family gathering.

It was, Terryl said, PTSD. Danny had taken Mitchell from her; she had willed herself away from that place, she'd moved far away. But she learned the limits of geographic distance; she carried with her the ghosts of Danny's abuse.

THEY'D MOVED TO LOGAN three years before that Mervyns incident, in 1995. Alan attended Utah State. After living in a condo, they'd found a house for sale in 1997 for $94,000. It was that cheap because it was infested with hornets, had no toilets, and was ill-configured. Of the three thousand square feet of livable space, fully half of it was the living room.

Weirdly, the house's lack of perfection appealed to Terryl. It let her feel like she was distancing herself from her mother, who Terryl always felt put on airs that everything was okay. Even though it was a big fat lie.

She decided to take a deliberately different approach. "I don't care if the house is perfect all the time." That would be superficial, she thought. It didn't mean she wanted a hornet-infested mess. They fixed it up, made it livable, just another long-term improvement project for Terryl.

She was taking these steps, making declarations, trying to rebuild herself just as she was remaking the house with the hornet's nests the size of basketballs.

During this period, she had another key rebuilding project on her hands. She was getting to know her real father.

That whole thing unfolded a few years earlier, when Jayme was eight months old and the family still lived in Southern California. Terryl and the baby were walking in a mall in Palm Springs with Kathie. A woman came up to Terryl's mom and said: "Are you Kathie Hartman?"

The woman was from Kathie's high school days. The woman asked Kathie how Woody was doing and whether Woody still worked at Kodak.

In an instant, it all came together for Terryl. Someone named Woody had been her actual father, and Michael's father, too. She'd never known her dad's name. Now she did.

At the urging of her friend Neal, Terryl contacted the Department of Motor Vehicles and managed to find this guy named Woody Hartman. She made contact with him and he told her that Kathie had told him that she was happy and had started another family. He says Kathie urged him to let her and the children move on with their lives, explaining his eventual absence, something he says he regrets. Sometimes, early on, Woody would call to see how the children were doing. When Terryl learned this, she thought she'd solved one of the mysteries of her childhood: Why she was told not to answer the phone.

"My dad would call and she didn't want me to pick up."

Terryl would occasionally burst into fury. "It breaks my heart," she says. "I had no father at graduations, milestones, my wedding.

"I lost out on a whole relationship in high school, middle school, college, without a father."

The years she lost with her father had Terryl thinking about being a victim from a different perspective. She developed a particular empathy for people who lost their moms or dads.

The more she learned about the accident on Valley View Drive on September 22, 2006, the more Terryl thought about the other victims—not just Jackie. There was Leila, and this eighteen-year-old girl, Megan, who was growing up without her dad.

ON JANUARY 2, 2007, Megan married her fiancé, Thomas Done. The ceremony took place at a small LDS church near the main downtown area. The bride wore a white gown with a top that bunched up around the middle. She walked down the aisle by herself, not wanting anyone to replace her dad. She finally felt the weight of his absence, after keeping it at a numb distance for many months.

"I couldn't stand the idea that my dad wasn't really there. I just didn't want to believe he was gone," she says of those months of denial.

At the wedding, on a brown upright piano in the church, she'd put a big picture of her dad. Her mother, she says, was crying hysterically. "It was like my dad's funeral all over again."

The marriage didn't work. Megan had been struggling for a few years—with school grades and swimming injuries, with her relationship with her parents. She hadn't gotten a job and that was one of several sources of growing tension with her new husband. They fought, really fought. Not a month after the marriage, the cops were called in after she and her husband got into a physical altercation. Megan was arrested. She got probation, a first-offender break.

Her life was getting away from her. She spent hours a day, sometimes all day, playing shooting games on the Xbox, either against her husband, or teaming up with him against other people they'd meet online. They'd fight with myriad virtual heavy weaponry across elaborate virtual terrain, seeing which team could get fifty kills first. She felt good when she was online, like she was skilled, and it was thrilling, the constant, intense interactivity. "I don't know how to explain it," she says of her passion for playing. "It's pretty much all we ever did."

AT THE START OF 2007, Megan and Leila were struggling to find anything close to closure around Keith's death. Jackie, with raw determination, was doing a bit better.

Terryl remained at arm's length—but not for long.

And Trooper Rindlisbacher was about to get a break.

CHAPTER 18

HUNT FOR JUSTICE

O N JANUARY 8, 2007, three and a half months after Reggie hit the Saturn and killed the rocket scientists, trooper Rindlisbacher's tenacity paid off: He got the okay to go after Reggie's cell phone records. It came in the form of an application to conduct a criminal investigation. It had been put together by Tony C. Baird, one of the county prosecutors.

On page twelve of the fourteen-page application, the document states that, just following the accident, Trooper Rindlisbacher drove Reggie to the hospital and "observed Shaw using his phone to receive and send text messages. The phone did not make any audible noise, but on several occasions Shaw pulled the phone from his coat pocket and sent a text message. Shaw held the phone with his right hand and used his right thumb to type and send each message."

A page later, it says that Trooper Rindlisbacher asked Reggie whether he'd been texting during the accident. "He denied using it. Shaw could not or would not give a reasonable explanation for his driving pattern."

But if it sounded like the district attorney's office was circling Reggie, that was far from the case. Baird, the prosecutor, had his skepticism—about what the facts would show and whether there was applicable law. For now, the document was a permission slip to Trooper Rindlisbacher. And the real meat of the document lay in its middle, pages six and seven. It was a subpoena, aimed at Verizon Wireless. The company was ordered to provide to the trooper, as an official of the state, records associated with the number 435-XXX–3739. Reggie's number.

"Copies of billing statements from September of 2006 until present. These copies should include both incoming and outgoing tolls of both calls and text messages. This should include all incoming and outgoing phone numbers dialed or received."

IN THE MIDDLE OF January, Leila got a call from Herm Olsen, the lawyer she'd previously contacted. There were two things on his mind. He'd been thinking about the accident and about the scene where it happened. "There's no shoulder on the road," he said. "The road isn't safe. Everybody knows it's not safe."

Leila thought back to what she'd seen at the accident site. On either side of the highway were about four inches of pavement, then a sharp drop into the gully. Olsen told Leila that it might be worth getting an engineer. He told her that, depending on what the engineer found, it could be worth suing the state.

"I don't care about that. I want the road fixed," Leila told Olsen. "Please, just get them to get the road fixed."

Leila recalls that Olsen took it in. He had one more thing to bring up with Leila: Had she heard what happened with the investigation? Was it over? Did Reggie get a ticket?

Olsen told Leila the police should have alerted her. But she told him that they hadn't. She resolved to find out. She hung up with him and placed a call to the county clerk's office. They hadn't heard a thing about it. She dug up the business card of one of the officers who had visited her after the accident.

Shortly thereafter, the phone rang. She sat down at the counter in her kitchen and picked up. "This is Bart Rindlisbacher," the man said. He explained that he'd been looking into what happened at the accident. Then he went into his story, about how he watched Reggie texting on the way to the hospital.

"What?" Leila was shocked. "This is so different from what I've been told."

The trooper told her that, despite getting permission to investigate, he was having trouble getting the cell phone records; in fact, the Shaw family wouldn't confirm which carrier was Reggie's, and the investigator was going company to company trying to figure out where to start. Leila had the impression that the trooper was open, and honest, but also dogged. "I really had the impression he was doing a great deal of this on his own."

She hung up and felt like she needed to do something, too. Texting and driving; could it really have happened? It seemed to her so absolutely absurd, so illogical, so dangerous. She thought back to Herm Olsen's firm. The first principal, Lyle Hillyard, was a state senator.

A seed was planted in the back of her mind, something tiny and still unrecognizable to her. Yet it would blossom. Leila would eventually need to do something positive with her grief. Maybe the senator could help.

CHAPTER 19

REGGIE

M ATCHUP ZONE! MATCHUP ZONE!"
Reggie, in a shirt and tie, standing next to a metal folding chair, shouted defensive instructions to five scrawny-armed sophomores of the Bear River Bears, Reggie's former high school team. It was winter 2007, during a basketball game against the Grizzlies from Logan High School.

It looked to be trouble. Earlier in the season, the Grizzlies had crushed the Bears. But things were tighter this game. Reggie was feeling the tension, getting lost in it.

Toward the end of the year, before Thanksgiving, he'd taken up an offer from Van Park, the varsity coach, to help coach the sophomores. Reggie was the assistant coach to Greg Madson, who doubled as coach and the publisher of the *Leader*, the Tremonton newspaper.

Madson couldn't make this game so Reggie was on his own.

An adage in sports is that "defense wins championships." It fit into Reggie's own personality on the court: hustle, do the unheralded jobs, let your teammates score. He was trying to impress this on the sophomores and, in this game, that meant constantly changing the defense to keep the other team off guard; man-to-man, and zone, 1–2–2, 1–3–1.

"Every other time down the court we were in something different," Reggie says. "It was just constant."

Bear River lost by six. He was initially disappointed, then felt some measure of moral victory, having kept it so close against the bigger school. Basketball brought momentary relief.

"In all honesty, the basketball was such a great release, an avenue for him to get lost," says Madson. He knew Reggie well, as a player and as a kid. Nearly in his fifties, Madson was in Reggie's church ward and had been an assistant to Van Park on Reggie's own varsity high school teams.

In the 2006–2007 season, just after the accident, Madson felt like Reggie mostly didn't betray the grief he felt. But then he'd see Reggie "in a quiet moment," and he'd see a "momentarily blank stare." Madson felt like Reggie was flashing back to the accident, but he didn't feel it was his place to pry. "You could tell it was eating him up."

Madson decided a good therapy might be to give Reggie more responsibility. And so the coach allowed Reggie more latitude to coach on his own, say, to take the guards for part of practice or the big men, responsibilities that he might not otherwise give to a first timer.

"When he spoke, they listened," and not just in the way that younger kids in that community did when coaches talked. "He'd had some real-world experiences. He'd been through some really tough stuff."

REGGIE WAS DATING AGAIN, his interest in women, if not at its full measure, returning. Her name was Trisha Haber. She was a year younger than Reggie and a good friend of Briana Bishop, the woman whom Reggie had been seeing at the time of the wreck.

"Our relationship wasn't the best before the accident. And after, it was kind of too much and it didn't work out," Reggie says of dating Briana.

Briana was from Kaysville, about forty-five minutes away. Reggie had met her through Dallas, his gregarious buddy. Trisha was one of the girls in the group they hung out with. She had curly hair and a dark tint to her skin, her dad being from Israel. Trisha worked in a T-Mobile store. She and Reggie went on dates, taking turns making the forty-five-minute drive.

She knew about the accident, seemed supportive, though they didn't talk much about it. They did discuss a future.

Marriage talk, not sex talk. They made out a couple of times. That was it. Reggie had learned his lesson, and he had a focused, overriding goal. He still would go on a mission. "It's what I wanted more than anything."

Maybe he'd be able to go in June, just pick up where he'd left off.

This is how it went in the spring of 2007: a kind of collective life-as-it-used-to-be with an unspoken feeling it could all burst apart. Leila O'Dell and her daughter, Jackie Furfaro and her daughters, and Reggie and his family began trying to reclaim their lives. There were fits and starts of normality, but always a specter loomed. Somewhere, out there, some machinery, embodied by the stubborn Trooper Rindlisbacher, was following a trail.

Occasionally, the Shaws would check in with Jon Bunderson. He had a stock cliché: No news is good news. And for his part, he wasn't about to contact Cache County officials. "Don't kick a sleeping dog," he described as his philosophy. He fielded some questions about insurance, from the Shaws and the insurance companies.

On the criminal front, there was a sense, in Reggie's camp, that maybe it would all go away.

Far from it. With the legal force of the Cache County Attorney's Office, Rindlisbacher had been faxing Verizon in late January and early February. The fax asked for phone records associated with Reggie's number. And the document included a copy of obituaries for Keith and Jim published in the *Herald Journal*, Logan's local newspaper, on September 25, three days after the accident.

It took at least five requests to get the records. And then, the trouble was, the information didn't make sense.

Rindlisbacher needed reinforcements.

A MEETING TOOK PLACE in the highway patrol office in Cache County, the same second-floor office where Rindlisbacher had come immediately after the accident. Sergeant Tony Hudson, who'd been working with him on the case, was there, along with a new guy, Scott Singleton. It was March 15, a Thursday, at 10:30 in the morning.

Scott was the new local agent for the Utah State Bureau of Investigation, assigned to the northernmost counties. Scott didn't much like his job's chief responsibility: hang out at bars and, as Singleton put it, "enforce Utah's crazy alcohol laws." A byzantine labyrinth of rules and regulations, he said, that he was supposed to enforce "sitting in a bar all night long, which I hated."

When he had time, though, there was a part of the job he loved. He was supposed to help pursue cases for the highway patrol that were taking a lot of time, that the individual troopers didn't have resources to pursue. Not cold cases, exactly, but complex ones.

The job suited Singleton after a long period wandering. Born in 1964 in the tiny farm town of Benjamin, Utah, the son of a plumber, Singleton was a nice kid but a poor student, shy, a follower, a practical joker. At Spanish Fork High School, where his grade point average didn't reach past 3.0, he was a reasonable class clown; once, when the National Guard came out to celebrate the school's newly renovated football field, Scott joined a few friends hoisting signs that read: MAKE FOOTBALL FIELDS, NOT WAR.

He kicked around at college or, rather, colleges: Southern Utah University, Utah Technical College, Weber State, Utah Valley. He didn't last more than a few semesters at any of these schools, finding himself driving with regularity to Arizona, where the booze laws were looser, partying regularly.

In the back of his mind, he started to wonder if he had a learning disability.

"I cannot pay attention for more than a few seconds," he says. "I'd sit in class, and no matter how hard I'd concentrate—if there's a window, I don't care where it is, I'm staring out of it—I'd have the hardest time paying attention."

He didn't get a degree, bouncing around jobs (farming, dry walling), got married at twenty-one, and took a job two years later working at the

Utah Port of Entry in Wendover, an entry-level law enforcement gig. He worked his way into a job as a road trooper, then a state investigator.

At the meeting that morning, Rindlisbacher handed Singleton two cases. They were Rindlisbacher specials, meaning: cases that he'd gotten personally invested in.

One involved a guy who randomly approached Rindlisbacher at a restaurant and offered to buy the trooper dinner. The guy was dressed as Santa Claus, and struck Rindlisbacher as "creepy," Singleton recalls. Then later, the same guy had been stopped while driving a taxi with a little boy in the back; Rindlisbacher did some checking and discovered the guy had plead guilty in Washington State to "immoral communication" with a minor.

Rindlisbacher "wanted to see if we could get him registered as a sex offender in Utah," Singleton explains.

Rindlisbacher handed Singleton the other file, Reggie's. He told the story to Singleton, who was taken aback.

"I didn't have a cell phone, I'd never texted. He might as well have been talking about nuclear physics."

In the file were several compact discs.

Rindlisbacher handed them to Singleton, and said: "I think we've got proof."

THE DISCS CONTAINED RECORDS from Verizon Wireless, the results of the subpoena.

They were not, on their face, especially helpful.

Two days later, Singleton sat down in a windowless room in Brigham City, his immodest investigator's surroundings, put one of the discs into the computer, and discovered what looked like nonsense. It was an endless list of calls and texts, intermingled, not organized by day or time. Not organized in any fashion at all—and spanning a period of months.

"It was just jumbled," Singleton recalls. He started trying to separate the calls from the texts, the long ones from the short ones. Then began organizing the data by date.

It took a few weeks, working on and off, in between patrolling the bars. He didn't have much contact with Rindlisbacher; there wasn't much to say.

Then, in mid-March, the data started to crystallize. Singleton was back at the computer, when something stood out.

6:47.

There was a text at 6:47 a.m., on September 22, 2006. The morning of the accident. *Wasn't that the moment of the accident?*

Singleton went back and looked at the crash report. He looked for Kaiserman's 911 call. When did that happen?

6:48.

And it was only moments later that Rindlisbacher, riding in the Crown Vic, had heard the call. Singleton shifted in his seat, cocked his head, tried to make sense of it.

"Holy cow," he muttered.

He played with the numbers some more, looked at the order of events, texts, and calls, which he'd finally gotten straight in the database.

Reggie had sent a text at 6:17.

Another at 6:43.

Another at 6:45.

Another at 6:46.

Singleton thought: *He was texting when the crash happened.*

He didn't call Rindlisbacher, not yet. He didn't want to be wrong. There were too many unknowns. Among them: Whom was Reggie texting? All the messages had gone to the same number: 801-XXX-3126.

Singleton considered calling the number. But then he thought, *I've got to proceed cautiously. Someone might answer and then just go cold on me, shut me down, or out.*

SEVERAL DAYS LATER, SINGLETON got out a piece of paper, and he wrote down the sequence of events from the crash report—when the accident took place, when the 911 calls came in—and he wrote down the times of all of Reggie's texts.

He climbed in the aging Smurf-blue Ford Taurus that was his state-provided car. It was 10:30 in the morning. He drove to Tremonton, Reggie's hometown. And then he commenced retracing Reggie's drive, matching times with texts, which had commenced with the one sent at 6:17.

As he approached the gun range, near the mile marker where the accident took place, Singleton was struck by two things: (1) he was having trouble watching the road and looking down at his piece of paper, a challenge that reinforced his astonishment that someone could simultaneously text and drive; and (2) there was no doubt: "Reggie had been texting the entire way. He had been texting when he caused the crash."

Questions whirred through Singleton's mind.

How do we prove this?

Whom was Reggie texting?

Who belonged to 801-XXX-3126, the number on the other end of the texts?

MARCH 17, 2007, WAS Saint Patrick's Day. In Salt Lake City, a seventeen-year-old girl named Lauren Mulkey had freshly returned from a family vacation in Florida. In the early evening, she put on a bright yellow halter top to show off her tan, jeans, and big green hoop earrings, and set out with friends to celebrate the Irish holiday.

The last thing she heard was a warning. "I told her: Be careful of drunk drivers," says her mother, Linda.

At midnight, now March 18, she was returning home, sober. Coming the other direction was Theodore Jorgensen, nineteen. At a thoroughfare near the University of Utah, he drove through a red light.

He did not see Lauren's Mercedes SUV passing through the intersection, perpendicular to him. He smashed into the driver's side of the car. It flipped. Lauren died almost instantly from massive head wounds. According to prosecutors, Jorgensen had been on the phone prior to the accident.

Lauren's mother, Linda, says: "I never thought to warn her about the phone."

CHAPTER 20

THE NEUROSCIENTISTS

I N A SMALL WINDOWLESS room in Fraser Hall at the University of Kansas, a junior named Maggie Biberstein sits in a contraption that resembles a driving video game at an arcade. There are brake and accelerator pedals at her feet. She grips a steering wheel. She looks at a screen on the wall on which is projected a highway scene, the setting she is supposed to steer a virtual car through.

This driving simulator is telling Dr. Atchley and his collaborators something about Maggie's relationship to technology.

Maggie's job is to drive the car and focus on the road. But she's also supposed to keep an ear out for her phone, in case a text comes in. That's not uncommon for her.

"I sometimes text while I drive," she says a bit sheepishly. She wears jeans and a green sweater over a blue shirt. She grew up in Manhattan, Kansas, where she got her first phone when she was sixteen. She trics not to use her phone when she studies, but it's hard to ignore. The device is so important for social uses; she's heavily involved in her sorority, Alpha Delta Pi, where, for instance, she recently helped write a musical.

What is the value of social connection? How does it impact the lure of the phone? And when it comes to social connection, What is the value of immediate gratification?

IN A PRIOR GROUNDBREAKING experiment published in 2012, Dr. Atchley and a colleague explored these questions. Dr. Atchley used a basic premise: He theorized that if technology was like an addictive substance, say, alcohol, then users would find the information delivered on a device as irresistible as alcoholics do booze.

In the first part of the experiment, thirty-five students were asked about how they value money and how they value information.

In the case of money, the subjects were asked if they would prefer to take a small amount of money, say, $5, versus waiting a period of time to get more money, say, $100. The students were asked about other time periods, too, and other amounts of money, enabling the researchers to assess how the students value money over time. Put another way: How urgent is their desire for money?

In the second part of the experiment, the researchers asked similar questions, but this time they focused on the value of information. For instance, students were told they've received a text from their significant other. They were promised they could get $5 if they responded immediately, or $100 if they waited an hour to respond. There were other intermediate options, such as respond in five minutes or thirty minutes with corresponding rewards.

This kind of science has been termed "neuroeconomics," a subset of which is known as "delay discounting." It is a way of understanding decision making on various issues—say, when a person responds to a text—by using economic, or monetary measures and influences.

What the researchers found was that students thought the value of a text fell much more quickly than did the value of money. On average, $100 lost about 25 percent of its value if a student had to wait 12 days to get paid, and it lost 50 percent of the value if students had to wait

142 days to get paid, on average. Despite some erosion, money retained value over relatively long periods.

Not so with a text. In the case of a text, the information lost one-quarter of its value in ten minutes and half its value in five hours. Information loses a lot of value in a short period of time. Money retains its value over time.

The results stand to reason, Dr. Atchley explains. "If a friend texts and says, 'I'll be at a party later,' and you don't pick up the text until tomorrow, you may as well not have bothered to pick up the phone."

In a second phase of the experiment, the researchers added a twist: Would the value of information be discounted even further—be even more urgent—depending on whom it was from?

What they found was there was a decided increase in the sense of urgency when the text came from a significant other, as opposed to a friend or more distant acquaintance. Specifically, when the text came from a boyfriend or wife, for instance, the value of texting back fell 25 percent after just twenty-five minutes, compared to three hours for a "friend," or ten hours for an acquaintance. The dry vernacular of the science paper read: "The current data shows the need to text now may simply reflect the need to engage in a behavior that only has value in the short term."

There was another critical conclusion from the experiment. The college students made what Dr. Atchley calls a "rational" assessment of the value of the information. They didn't just have a knee-jerk reaction to immediately respond to the text. If they had done that, Dr. Atchley thought, their relationship to crucial social information would begin to look more like that of an alcoholic to booze. In that relationship, the alcoholic places urgent value on, say, getting one beer today, even if he can get a twelve-pack if he waits a week.

The college students, by contrast, were willing to wait a bit, and they were prepared to wait more or less time depending on whom the text was from. That is a rational response, Dr. Atchley thought, not the response of an addict.

Maybe, he thought, technology creates more of a compulsion than an addiction. He needed to dig deeper to understand where the lure of technology fit on this continuum.

IN ANOTHER EXPERIMENT AIMED at refining the answers, Dr. Atchley is working with scientists at the Imaging Center in Kansas City. There, they are comparing the brains of smokers to those of Internet addicts. When a smoker anticipates a cigarette, does his brain look the same as when a Facebook user anticipates getting a status update?

And in yet another experiment, his team is asking a question related to distracted driving. What is more powerful: the lure of the social information or the demands of the road?

That's what Maggie is doing in the lab today. It's just one little building block, but they're about to be surprised by the result.

IN THE ROOM NEXT to where Maggie sits in the driving simulator is a small room with three computer screens and one television monitor. These are receiving a feed from a camera that's recording what Maggie is doing.

Watching is Chelsie Hadlock, a researcher in Dr. Atchley's lab. As she observes Maggie, she uses one of the computers in the room to send Maggie occasional texts. There are two kinds of texts: one involves driving directions that Maggie must take to get to a party; the other are related to social information about the party and who will be there.

The researcher taps out: *Michelle keeps asking when you're going to get here.*

Maggie taps back: *I think I'll be there soon. How do I get there?*

The researcher texts how to get to the party, what street to turn left on, and how many blocks Maggie must go. Maggie drives on through the simulator, finishing the highway portion, heading into an urban environment.

The researcher texts: *Gendry is getting wasted. Why aren't you here yet?*

The simulation is not intended to measure how well Maggie texts and drives, though the researchers confirm that she and other participants are horrible at it. Crash risk soars. Something else happens, too, perhaps less expected; the social information seems to pave over the value of the driving information.

At the end of the simulation, Maggie takes a quiz. What does she remember about the drive? What she recalls are the names and details of the lives of all the fictional characters from the party: Michelle and Gendry and Michael.

What she misses in the quiz: everything else. The driving directions, the number of intersections she passed, the buildings she passed.

GETTING INFORMATION OF VALUE seems to explain some of the powerful lure of technology. Here, it is possible to see the way in which our devices play so beautifully to our two basic attention systems: top-down and bottom-up. Our top-down, goal-directed system wants to keep in touch, make connections, form relationships, forge partnerships. Our devices are masterful at allowing that. And the goals get reinforced, or so it seems, by the buzz of an incoming call or text, alerting us to a new development in the narrative of our lives. It is the bottom-up system at work.

But there's another way in which our devices, perhaps less expectedly, cater to our deepest neurological wiring. It has to do with the value of *sharing* information, not just receiving it. Turns out, we also get important neurochemical releases when we disclose personal information, something enabled greatly by email, texts, or status updates on various sites.

In a study published in the *Proceedings of the National Academy of Sciences* in 2012, researchers from Harvard University used MRI scans to look inside the brain and ask: What happens when we disclose information?

What they found was that the reward areas of the brain light up when people share. "Here, we suggest that humans so willingly self-disclose because doing so represents an event with intrinsic value, in the same way as with primary awards such as food and sex."

Reinforcing that finding is research by Dr. Gloria Mark, a professor at the University of California at Irvine, who found through survey studies that people are happier at work when they use Facebook more. Not more engaged with their job, but happier. That comes in part from posting stuff, sharing.

The researchers added another layer to that finding by looking at whether the reward circuitry in the brain was more or less excited when the person was disclosing the information not in a vacuum or in private—say, by writing the information onto a piece of paper (a diary, if you will)—but when the disclosure was certain to be seen by someone, particularly a friend or family member. To measure this, the researchers conducted a version of the neuroeconomics experiments that Dr. Atchley (and many others) had used, in this case trying to put a relative monetary value on disclosing something in private versus disclosing to another person.

"Although participants were willing to forgo money merely to introspect about the self and doing so was sufficient to engage brain regions associated with the rewarding outcomes, *these effects were magnified by knowledge that one's thoughts would be communicated to another person*, suggesting that individuals find opportunities to disclose their own thoughts to others to be especially rewarding," the researchers wrote (italics added).

They concluded: The "motivation to disclose our internal thoughts and knowledge to others may serve to sustain the behaviors that underlie the extreme sociality of our species."

PART OF THE REASON why it's not so easy to understand the powerful lure of technology—is it addictive? the cause of extreme compulsion? or simply habit forming?—has to do with the complexity of the technology itself. Various mechanisms are at work. Some have to do with the nature of the information the devices deliver—the value, say, of personal information or the value of good news versus bad news. Some are more purely mechanical, having to do with *how* the information is delivered.

For instance, the speed at which a device delivers information can dictate how drawn we are to it. Another mechanical piece has to do with the way our brains get stimulated by the mere act of touching the device and, in doing so, causing something to happen. Touch the keyboard and a letter appears. Touch the screen and an email opens or a gun shoots in a video game. This step is separate from, and arguably more primitive than, the receiving of information; before we get information, we are getting some stimulation through an act of stimulus response.

An additional factor that can dictate how compulsively we use our devices has to do with individual personality and predilection. Some people appear more susceptible than others.

Different researchers are exploring the different mechanisms, and they are doing so using a growing set of tools. Some research focuses on behavior, like the stuff Dr. Atchley does. Some involves imaging, Dr. Atchley does that, too, but it is more the domain of scientists like Dr. Gazzaley. And there's another way of looking at these various mechanisms. It entails looking at neurochemicals triggered when people play video games or use the Internet. The tools to measure such activity are evolving, too. And they show that the use of certain kinds of interactive media stimulate neurological patterns similar to those produced by the use of addictive drugs.

To learn more, Dr. Atchley suggests I talk to an Internet addiction specialist named David Greenfield. He has looked closely at the chemical relationship between drugs and technology. And he knows whereof he speaks, having struggled with drug use himself.

CHAPTER 21

TERRYL

MARY SURRATT RAN THE boardinghouse where John Wilkes Booth planned the assassination of Abraham Lincoln. Evidence of her guilt was conflicting, but she was found guilty of conspiracy in the president's murder and sentenced to hang.

Early in 2007, as the case against Reggie was inching along, Terryl's oldest daughter Jayme, a sixth grader, had become interested in Mary Surratt. Jayme started building an exhibit for the Utah History Fair. It included a description, photos, and a hangman's noose. What made the noose unusual was that it had five loops in it, rather than thirteen, just like the noose that Surratt's hangman had fashioned at the time because he didn't believe Surratt should hang and he thought a noose with fewer loops would break.

In early April, Jayme won the junior division of the state competition, making her eligible to compete in Washington, D.C., on National History Day that summer. It was the first of what would become a string of academic successes for the Warner children, in the classroom and in competitions outside of it.

Jayme formed the early idea she might be a doctor. "I hate to see anyone in pain, or suffer." She excelled in school, setting a bar her brother felt eager to match. When Taylor was in the second grade, the school's principal suggested Taylor skip third grade and move on to fourth. That year, in fourth grade, Taylor went to the house of a friend whose dad was a heart surgeon. He decided right then and there he wanted to be a doctor, though not a heart surgeon because, he thought, "Ooh, the heart's gross." He became determined to be a neurosurgeon. "The brain," he says he'd concluded, "is so cool."

For Terryl, the exhilaration around learning was a sign of the enthusiasm she tried to instill in her children, tried to exhibit for them. This was the personal analogue for Terryl of her zeal at work. She was desperate to be an engaged parent, 180 degrees of difference from how she felt her parents treated her. If a child got interested in a project or hobby, Terryl was there, driving them to whatever event, staying up late with preparations. It was the zeal of a convert and, just as at work, her enthusiasm at home could tread a line.

In the back of the house, she and Alan had cleared out a big backyard. But it wasn't particularly hospitable to play. Then it sunk in with Terryl that the neighbor had a backhoe.

"I decided I wanted a sandpit for the kids."

With the neighbor's help, she dug a hole, around four feet deep, twenty feet wide, and ten feet long. She ordered a huge dump truck full of sand, and she had it poured into the hole. There's a picture of Taylor with his friend and neighbor Travis, both five years old, buried up to their necks in sand.

They had a chocolate Lab named Luke. Jayme and Taylor would tie a red wagon to Luke and then throw a ball across the sandpit and the dog would race to get it, the wagon trailing behind. In April of 2001, Allyssa was born, and she was little more than one year old when her older siblings would put her in the back of the wagon and Luke the dog would pull her around, sometimes chasing the ball.

Terryl and Alan put in a tetherball court, a trampoline, a swing set. Money was tight, but it was low-cost entertainment. Terryl could garden and be with them without being in their faces.

Sometimes, Allyssa would fall out of the red wagon, with Luke the

Lab carrying her across the sand. "She'd go flying out," Terryl laughs, looking back. The girl didn't get hurt. But it did exemplify how Terryl, in her eagerness to be supportive and fun, to encourage a spirit of adventure, could push things to the edge.

When Taylor was around six, he expressed an interest in science. Terryl got him a microscope and a chemical experiment kit that had test tubes and glass straws, an alcohol burner, steel wool, and small containers of some not insignificant chemicals, like hydrochloric acid, sodium carbonate, and calcium hydroxide.

"My mother was always there and always giving us crazy things to do," Taylor says.

A few years later, she took them to a nearby park, a nearly daily activity. Taylor wanted to ride the zip line. It was this thing where the adult gave you a little shove and you glided about eight feet to the other end of the line, feet dangling a few feet above the ground.

"I asked my mom to push me," Taylor recalls. He rode the line. "I asked her to push me harder," and she did. "I asked her to push me really, really, really hard, and she did, and I fell and I broke both my wrists. It was funny, but not at the time. At the time it hurt."

It was a hard line for Terryl to draw—the line between support and engagement and going overboard—wanting so much to be involved and to let them have a childhood. Always there was the specter of what she'd missed. "I lived it every day," she says, thinking back.

"The way I raised my children, I put a stop to domestic violence in my home. I put a stop to alcoholism in my home. I was a chain breaker, is the way I view it."

THERE WAS ONE STEP in particular she felt was essential in spurring her children to be engaged at home and in school. When Jayme was five years old and Taylor was three, Terryl cut the family's cable TV service. Lots of families cut television when the kids are young, hoping to limit screen time. For Terryl, the step was substantive and symbolic; one of the few things she felt was good about her own childhood—other than her devotion to the church and reading—was that she didn't watch a lot of television. Her mom sent her and Michael out of the house, into the yard, which Terryl thought had less to do with any enlightened parenting philosophy and more to do with keeping Danny happy.

There was another motivator; Terryl remembered her mother-in-law, Alan's mother, as someone who "didn't get out of bed, because all she wanted to do was watch TV."

"We were both greatly affected by parents who didn't get involved," Terryl says. "I just wasn't going to plop them in front of the TV."

TERRYL HADN'T LOOKED AT the scientific research, but it was showing that television could have a powerful impact on the relationship between parents and their children. That's because television plays to our attention systems in extraordinary ways, using light, sound, and story. The science explains why, for all the growth in our other gadgets, television remains the single most dominant form of media.

WHEN THE TELEVISION IS on, parents and children disengage from one another. The parents, even if not instructed to watch television, talk less to their children and respond less to children's inquiries and efforts to get attention, according to a summary published in 2009 by some of the field's leading researchers at the University of Massachusetts at Amherst.

In the study, parents interacted with their children 68 percent of the time when the television was off and 54 percent of the time when it was on. Further, the research showed, the "quality" of the interaction fell, too, with the parent less likely to be engaged or even look at the child when they do interact.

But so what? So what if parent-child interaction falls? Researchers said that when parents talk to their children less, engage less—in a nutshell, put their attention to the television not the children—it can eventually retard language development. As the 2009 study concluded: "The evidence is growing that very early exposure to television is associated with negative developmental outcomes."

It also appears young children who watch television can experience shorter-term effects. In 2011, *Pediatrics*, the scientific journal of the American Academy of Pediatrics, published the results of a study that measured performance of four-year-olds on certain mental tasks after showing some of them nine-minute clips of a fast-paced cartoon, *SpongeBob SquarePants*, while showing other children either a slower-paced show or having them draw and not watch television at all.

The children who watched the fast-paced show were less able to follow directions and, in a separate set of tasks, showed less patience. These are "executive function" tasks, meaning they engage the prefrontal cortex, that all-important part of the brain involved in focus.

The researchers wrote that the toll taken on executive function came not just from the fast pace but perhaps from the fantastical nature of the cartoon, which gave the children's brains a lot of information to digest, thus potentially depleting cognitive resources. The researchers wrote: "The result is consistent with others showing long-term negative associations between entertainment television and attention." Among those studies, one published in *Pediatrics* in 2004 found that children who watch more television in their toddler years are significantly more likely to have attention problems by age seven.

There is another impact from high television use. Heavy television watching creates a sedentary lifestyle, less activity, more weight gain, even obesity. The weight gain comes not only as a result of the sedentary nature of television watching, or even chiefly because of it. Rather, it's the advertisements for junk food, notes the Harvard School of Public Health in a health advisory titled "The Small Screen Looms Large in the Obesity Epidemic."

"Increasingly, though, there's evidence that watching TV—and, especially, watching junk food ads on TV—promotes obesity by changing mainly what and how much people eat, less so by changing how much they move."

SO WHY IS TELEVISION so alluring? The reasons are more complicated than meets the eye. They have to do with attention capture, and attention theory. One of the leading researchers in the field, who helped write the 2009 Amherst paper, is Daniel Anderson, a professor at UMass Amherst, and he's anything but a purebred television antagonist. In fact, he's consulted on numerous television shows, including *Sesame Street*, *Dora the Explorer,* and *Blue's Clues.*

Dr. Anderson has concluded that television's lure owes its powerful appeal to two parts of attention that often compete: the top-down system and the bottom-up system.

To recap: Top-down attention is the power people have to set goals and focus on them, while the bottom-up attention systems—very primitive systems—get captured by changes in light and sound and movement and, in effect, warn the top-down systems that there is a potentially relevant change to the environment, perhaps an opportunity or threat.

Dr. Anderson has found that babies mostly glance at television because of the bottom-up system—in other words, they are responding to changes in sound and light. "When there is a change, you want to find out what it is."

Then, as early as 1.5 years old, children begin to process the information on the screen, and then a higher order of thinking gets involved, Dr. Anderson has found. They want to understand what's happening on the screen. This begins the engagement of their top-down attention systems, as the children focus on what's happening in the show—the content—rather than merely being drawn in by the lights and sounds.

This doesn't make the changes in light and sound and movement irrelevant. In fact, what makes television so powerful, Dr. Anderson notes, is that it uses these attention-capture tools to constantly draw people back into the story, should their top-down focus waver. "Television always had that lower level working for it," he says, noting the mix of top-down and bottom-up tools: "Well-produced television makes it a fairly seamless experience. You're getting attention brought back to the screen in a way you don't feel you're being manipulated."

Adding to the lure, he said, there is a concept of "attentional inertia," which shows that the longer someone stays engaged with the television, the more likely that person is to have trouble pulling away. This is the case even when the television flashes to a commercial, Dr. Anderson explains, suggesting that the attention systems lock on even when the narrative isn't engaging the top-down systems.

Dr. Anderson has worked on many of the most popular children's shows, and is far less critical of television than some researchers. He thinks the right programming can be educational and informative. At the same time, he agrees there is reason to be wary of too much television viewing, largely in light of the opportunity costs. The more you watch television, the less you're interacting with other people, talking to your parents, moving your body.

"It doesn't make watching television bad for the kids," he says, but adds: "There are probably a thousand better things to do any given time than watching any given television program."

Television turns out to be powerful because it so effectively plays the top-down and bottom-up attention systems against each other. But that is mere child's play compared to the way personal communications devices can commandeer those systems. Television, after all, has plots that, while interesting, are not personal. Sure, you care what happens in *Orange Is the New Black* or *The Real Housewives of New Jersey*. But your phone and computer are bringing you a plot that is about *your* life. Your top-down system is heavily invested, and it gets reinforced by the bottom-up pings alerting you to what are, in a sense, plot updates. And so, as powerful a draw as television remains, our modern devices command our attention like nothing before.

CHAPTER 22

HUNT FOR JUSTICE

SINCE REGGIE WAS A kid, he'd had a few favorite passages from the Book of Mormon. After the accident, another passage spoke to him: "Therefore this life became a probationary state; a time to prepare to meet God; a time to prepare for that endless state which has been spoken of by us, which is after the resurrection of the dead."

Throughout the spring, the family worked through a process that was both bureaucratic and spiritual, aimed at getting Reggie back on track for a mission. He had more than a few hurdles, given that he'd already returned home after lying to a bishop about his premarital sex. And now he had this amorphous, ominous, legal cloud.

In April, church leaders explained the steps he'd need to take. Among them was composing a letter about his interests and commitment. He wrote that the tests of the last year had brought him closer to God and to his religion; that he'd never wavered in his faith; and that, but for a few months when he'd been absent from prayer and the church, he'd found time to do his studying and reading.

"Even on my worst day, a mission was something that I knew I wanted to do."

In this unusual case, Mormon officials needed to get more information. Records show that Jon Bunderson, the Shaw's attorney, talked to church officials in early June, giving them qualified reassurance that Reggie could go on a mission. He'd had no indication otherwise from police investigators that Reggie was at fault in the deaths of Jim and Keith.

SCOTT SINGLETON'S CASE NOTES show that from late March through April and into May, he hunted down whom Reggie had been texting on the morning of the accident.

If he could find out who the number belonged to, he could try to prove what the phone records seemed to show: that Reggie was texting during the accident, or right around it. He could find out what Reggie had been texting about; what had distracted him; was there a much clearer, if tragic, reason that two men had been killed?

Singleton's frustrations in trying to track down the answers were understandable to anyone who has ever had to deal with the phone company—navigating phone trees or getting help from people on the other end of the line who don't know the answers. There was an added layer of complexity when it was a legal case. The phone company had to make sure the subpoena was in order. Now there were multiple phone companies and different numbers. Lots of bureaucracy.

On March 19, Singleton met with Tony Baird, the deputy county attorney, to get a subpoena on the new, mystery phone number that appears on Reggie's records as being texted at the time of the accident.

For his part, Baird continued to keep the case at an intellectual arm's length. It was still very early in the process. He had plenty of work to do, a newborn at home—the fourth for Baird and his wife. Besides, Baird remained skeptical that the Shaw case would amount to much. He saw his job at this point as mainly making sure that Singleton was crossing his t's and dotting his i's. And he was impressed by this new investigator: "He was meticulous. He was very, very particular. Not that Rindlisbacher isn't," Baird recalls. "Singleton's more analytical, step by step, block by block."

By now, though, Singleton was also doing his best Rindlisbacher impersonation. He was obsessed. Not only did he get everything signed by Baird that March morning, he also dropped the subpoena off at 444 North Main Street, the Cingular store, to a woman identified in his case notes only as "Kris."

Exactly two weeks later, at 3:20 in the afternoon, Singleton called the Cache County Attorney's Office to see if they'd heard back from AT&T. No, the company had not complied. So the next day, Singleton met with Baird, who suggested the investigator call AT&T's legal department. Adding yet another wrinkle, Cingular and AT&T had recently merged.

He wound up calling Cingular's legal department, who told him such requests could take ten to fifteen days, an amount of time that had already elapsed. Singleton was getting frustrated, and he wasn't the only one. On April 13, a Friday, Singleton heard from Rindlisbacher.

"Keith's wife, Leila, wants to know what's going on with the subpoena," the trooper told the investigator.

"Working on it. I think I'm close."

Singleton wasn't. There was a practically comical twist coming.

On May 7, Singleton called AT&T's legal department, another pester. His case notes read: "They stated that the phone # belongs to T-Mobile."

He'd been barking up the wrong phone company. All this time wasted! But a few hours later, after talking to T-Mobile's legal department, he came to understand how he could make such a mistake. T-Mobile confirmed that the phone number had switched to its service, from Cingular/AT&T, on September 6. That was just three weeks before the accident.

That very day, Singleton wrote another subpoena, took it to Baird on Tuesday, May 8, and got it off to T-Mobile.

Then, not surprisingly, more bureaucracy, waiting, and back-and-forth.

On May 23, he spoke to T-Mobile and was told he needed to fax the subpoena to a legal compliance agent in T-Mobile's Law Enforcement Relations Group. That same day, at just a few minutes after noon, he called Leila to let her know the gears continued to grind. The next day, Singleton pestered the T-Mobile agent.

She told Singleton she had news. "We're faxing the information tomorrow." When he heard it, he felt a chill.

HIS CASE NOTES DON'T reflect the excitement he felt when the fax arrived.

"Received fax from T-Mobile. Briana Bishop, 12/8/87, is the owner of the phone # (801) XXX–3126."

Briana Bishop, age nineteen. The woman on the other end of the text.

THE DAY AFTER RECEIVING the information about Bishop, Singleton called a fellow investigator, Stan Olsen. Previously, Olsen had held Singleton's post in Brigham City but had since moved to the Farmington office, which was located closer to where Briana Bishop lived. Singleton and Olsen discussed how to approach Briana, and they decided to jointly interview her after Singleton made the initial contact.

He called the mystery woman at 2:30 on May 29.

"Ms. Bishop, my name is Agent Scott Singleton. I'm calling from the Utah Bureau of Investigation."

"Hello. What's this about?"

"I can't tell you over the phone. I'd like to arrange a time to get together."

At 2:48, Briana called back. They arranged to meet the next night after work at the Farmington office. Singleton was excited. He might finally get some concrete answers.

It wouldn't be so easy. Between the time that Singleton called Briana at 2:30 p.m. and the time she called back, she had two other phone calls. One was made to her friend Trisha, who was dating Reggie. The other call was with Reggie himself. It lasted eleven minutes.

THE NEXT NIGHT, SINGLETON and Agent Olsen gathered around a wooden table in the modest Davis County office of the Highway

Patrol in Farmington with Briana Bishop and her father, Steve Bishop.

In Briana, Singleton saw a "young, blond, nineteen-year-old who was nervous, very nervous." She wore work clothes. Singleton wore a dark shirt and a tie. He recorded the interview.

"You're not in any trouble, but we do believe that you have some critical information that you can supply to us," Singleton told her, according to the transcript. "I'm going to show you a series of photographs," he said. He showed her photos of young men, white, short hair, clean-cut, young men who at a distance could be mistaken for Reggie. She didn't recognize them.

He showed her Reggie's picture. "Do you know that young man?"

"Yes."

"Okay. And who is he?"

"Reggie Shaw."

Singleton asked their relationship and she explained, "Umm, we kind of dated a little bit, but we were mostly just friends. We hardly talk anymore."

Singleton asked what she knew about the accident.

"A little bit, not too much, though," Briana said.

"What can you tell me about it?"

"Umm, the morning that it happened, I got up, and I was getting ready for work. And I just texted him like I normally did, because I was dating him that day. And I so texted him to say good morning and what was up and stuff. And he texted me back and told me that he had just gotten in an accident and he thinks that the other two guys were dead and he was freaking out."

"And so he texted you and said—"

Briana cut him off. "It was after it happened. Then I called to see if he was okay."

Singleton had figured this would be a tough interview and that Briana, being a friend of Reggie's, may be evasive or even dishonest. He was also extremely aware of the other key person in the room, Briana's dad. As Singleton would say later: "We were having to appease her father. He didn't want to see her raked over the coals." Initially, Singleton started to try to draw her out. He pressed gently for her to explain the initial text. Who had sent it? She had, she said.

And then Agent Olsen asked: "You said good morning and his response was . . . ?"

"I just got in an accident and I'm freaking out and I think the other two guys are dead."

What Briana apparently didn't know is that the investigators had all the texting records. Singleton asked: "Do you think it's possible maybe the accident occurred when he was reading your text?"

"I don't think so, because he told me right after, he said that he had just—he's not one to sit there and text and stuff while he drives. He'll hardly like—he'd hardly like even pick up the phone or anything while he's driving."

Singleton's frustration was simmering. He pulled out the phone logs.

"These are a list of text messages. They start—the crash happened at 6:49. The texting starts prior to that time, 6:43, 6:45, 6:46, 6:47. These were all happening when he was driving."

Singleton had gotten the crash time slightly wrong—it was more likely between 6:47 and 6:48—but he made his point.

Briana: "Hmm."

Singleton elaborated, explaining that the texting started when Reggie left his house and that the investigation shows he "was in the process of texting you, and he crossed the center line and he clipped a . . . and the car spun out of control, spun sideways, and was hit by a pickup truck and it killed these two individuals."

"Uh-huh."

Singleton pulled out pictures of Keith and Jim that ran with their obituaries in the local paper. "This is a father who has a daughter, and this one is a father who has two little children. They were both married. I would really like as much help as I could because Reggie refuses to talk to us."

"I thought that he just texted me right after it happened."

Singleton said he had records for eleven texts between Reggie and Briana prior to the crash and for thirteen texts afterward.

And her dad then asked a question. "So these are responses—she sends a response and he sends one back, that's considered a text message?"

Yep, Singleton explained. He was trying not to show his frustration. This young woman remembered sending a text to say good morning, something so insignificant, and claimed not to remember anything else. "The only one I remember is just like the one after the crash happened, like he just texted me. That's all I remember about it."

They went back and forth, gentle parrying, trying to draw her out. Her dad said: "I think it's important you tell them everything you know."

"I know."

"Every little detail."

"Like, that's seriously all I remember about it. It was, you know, last September or so."

This was just shy of halfway through the interview. The investigators quickly covered a bunch of other ground: Who else Briana had talked to about the accident; how many texts she had sent that day; a bit about the process going forward. Briana asked: "So are you guys trying to get it so he goes to jail?"

"What we want is accountability," Agent Olsen responded. "If someone were to be texting and clipped your father while he was driving and was—and your father was killed," he asked, "would you want accountability?"

She had nothing further to offer—an indication of a standoff moving to a new phase, into a more formal legal arena, with the sides entrenching.

Indeed, by now, Singleton and Rindlisbacher, emboldened by discovery, were nowhere near close to giving up. In fact, Briana's own texting and phone records had offered them yet another level of indignation. Her records showed that she had sent Reggie six texts during his drive; he'd sent five, and she'd sent six.

Eleven texts, sent and read by Reggie Shaw on his fateful drive.

But Reggie didn't know anything of this discovery. He felt in the clear. Besides, something very good was about to happen, or so it seemed.

REGGIE WAS IN THE living room when the call came in. It was the local bishop, Eldon Peterson. He'd taken over for David Lasley, the bishop who Reggie had lied to about having relations with his girlfriend.

Reggie took the call on the landline.

"Great news, Reggie."

Before the next words came out, Reggie's eyes had already filled with tears.

"You're going on a mission," Peterson told him. The bishop gave a few key details, namely, that Reggie would be leaving in less than two weeks. Get your bags packed, kid, you're heading to Provo to the MTC, the Missionary Training Center, then to Canada, where you were supposed to go on the first mission.

"I know how hard you've worked, Reggie."

Reggie wept with relief and joy.

ON JUNE 7, AT three p.m., agents Singleton and Olsen took another crack at Briana. This time, unannounced. They showed up at Bukoos, a freight warehouse where she worked. They recorded the interview sitting in a back room with cubicles in it. They were the only people in the room but they shared a cubicle, the three of them, doing the interview.

Singleton was sure that Briana wasn't telling them everything, and suspected she'd been coached, maybe by Reggie. By coming to her place of work, they could avoid having her dad in the room and press a little harder.

Singleton explained that the prosecutors were now looking at the case. "In pouring through the records, text and cell phone records, I got quite a bit of evidence. I know you haven't been totally up front with me on what was said."

"I haven't?" Briana asked.

Singleton again explained about the cell phone records, the eleven text messages that took place before or just during the accident. And Briana responded: "That's what me and my dad were talking about after we left, you know, because like I thought—I told my dad like I thought I had only text him once, but apparently I was texting him before. I don't know."

"So you're saying you could possibly have been wrong?"

"I could have been wrong, yeah. I wasn't completely sure."

Singleton was feeling a light buzz, the thrill of closing in on some truth, something that had been elusive. He followed up by asking about her exchanges with Reggie after the accident; what had Reggie said? She explained, as in the previous interview, that he'd gotten into an accident. "They just ran into each other."

Singleton: "I want to know absolutely everything."

"I don't know what else to tell you. That's like honestly all I remember."

Singleton's frustration returned.

"He's looking at possibly a year at the maximum," the agent said.

"Uh-huh."

"An obstruction of justice charge is a felony. An obstruction of justice is when you have information and you don't provide that information."

"I'm telling you honestly like everything that I know."

The investigators turned their attention to before the crash. "So before the accident were you two texting back and forth to each other?" Olsen asked.

"I guess, if it says we were."

"It shows 6:17 in the morning and there's eleven texts."

"So I guess we were, yeah," Briana answered.

The investigators could see there would be no grand admission, beyond Briana acknowledging the existence of the call records. She wasn't going to say that Reggie admitted to her what he'd been doing, or that he'd been doing it at the exact time of the wreck. At that point in the interview, Briana's manager interrupted and asked how much longer the interview would take. The investigators began to wind down. They had a few more questions. Singleton asked Briana about a phone call that took place just a week earlier, the night before their previous interview of her. He had gotten the phone records and he could see that Briana and Reggie had spoken. He wanted to know if Reggie had coached her.

"Did he give you any indication of how to answer questions or don't tell them this or that?"

"No, he never said that. He just said like answer honestly is all he said. I don't think he's trying to hide anything."

Singleton left disappointed. He knew the prosecutors were on the fence, at best, about charging Reggie with negligent homicide. It would've helped if Briana could have said something to the effect that Reggie had confessed culpability, or said something about texting, or even coached her to obstruct the investigation. "I was hoping there would be some kind of smoking gun. But it didn't exist."

ON JUNE 11, MARY Jane called Bunderson and told him that Reggie was planning to leave in ten days for the Mission Training Center, then on to Winnipeg. Bunderson put in a message to Michael Glauser, the attorney for the Church, seeking guidance. Reggie was on his way.

CHAPTER 23

THE LAWMAKERS

IN MAY OF 2007, as Singleton hunted for answers, the governor of the state of Washington, Christine Gregoire, signed into effect the first state law banning texting while driving. She was, according to the *Seattle Times*, "flanked by children who suffered serious injuries after being hit by drivers."

The article noted that the texting ban, which would go into effect the following January, would carry a $101 fine for violators, while a second measure, also signed by the governor, would ban motorists from using a handheld phone. That law would go into effect in July 2008.

Across the country, legislators were grappling with whether to regulate cell phone use by drivers. The most intriguing battle had gone on in California. There, a state legislator named Joe Simitian, whose district included Palo Alto, the heart of Silicon Valley, had been nearly beside himself since 2001 as he battled the cell phone industry over a proposed ban on handheld phone use by drivers.

When he'd first introduced legislation, he stood in front of a legislative committee and explained that all he was asking for was to codify into law something that, he said, the cell phone companies already acknowledged: Using a handheld cell phone while driving was dangerous.

For instance, in an early hearing at the statehouse, in April 2001, Simitian, a Democratic assemblyman (later elected to the state senate), read from Sprint's own marketing materials. "When using your Sprint PCS phone in the car, focus on driving, not talking, and use your hands-free kit," the document read. "Failure to follow these instructions may lead to serious personal injury and possibly property damage."

As Simitian stood at the podium, reading the document, trying to reconcile the company's clear recognition of the risks with its opposition to law, he said: "I am at an absolute loss."

But year after year, his legislation had been killed. Lobbyists from the major cell phone companies, including Cingular, Sprint, and AT&T, threw a kitchen sink of arguments against the rule, arguing, among other things, that mobile phones posed no different a distraction than other tasks, like eating. Lobbyists for Sprint argued that a law banning handheld phone use could be used to discriminate against minority motorists, who were more likely to be pulled over because of racial bias by police. Broadly, the carriers argued that general education was sufficient to remind motorists about the risks of getting distracted by their devices. Never mind that cell phone use by drivers was exploding.

In fact, Simitian pointed out that the California Highway Patrol had been collecting data since 2001 and found that cell phone use was the number one cause each year in distractions leading to car accidents.

In 2006, after five years of battling, he got his bill passed. It was signed by the governor in September of that year, and was slated to take effect in July 2008 (to give law enforcement, consumers, and companies time to adapt).

Next up for Simitian was a texting ban.

But the idea of Utah getting in on the action seemed unlikely and even preposterous. This was, after all, a deeply red state, one that by numerous measures was among the top five most conservative states in the country. State legislators did not look kindly on government regulations they felt would impinge on personal freedoms.

By way of example, there was no primary law requiring the use of seat belts by drivers, and no law requiring motorcyclists to wear helmets. About the time legislators in other states were thinking about restricting bans on cell phones for motorists, Utah legislators were busy rejecting a different safety measure to require booster seats for children up to the age of eight years old, recalls Carl Wimmer, a Utah policeman turned state lawmaker from a Salt Lake City suburb and one of the more conservative members of the state legislature. And he sat on the Law Enforcement and Criminal Justice Committee, which was a gateway for these sorts of legislation.

Wimmer recalls safety advocates and families coming to the hearings. "They'd line up these kids who would come up and say: Please help save my life," he says. Wimmer felt for them, to the point that he at one point put up $1,000 of his own money to buy booster seats and give them away to any family who wanted one. But he wouldn't vote for such a law, arguing that it was just another example of the government sticking its nose into people's business.

He felt the same about texting and driving bans.

"If you're going to live in a free society, you have to give people the liberty to do what they want."

CHAPTER 24

THE NEUROSCIENTISTS

IN APRIL 2012, DR. Atchley attended a traffic safety conference in Orlando, Florida. He found himself on a panel with an Internet addiction expert with a particularly intimate knowledge of addiction, having spent time himself in rehab.

His name: David Greenfield. In the early 1970s, at Paramus High School in northern New Jersey, when Greenfield was just shy of fifteen years old, he was called to the principal's office and given a choice: Go to rehab or get expelled.

It was post-Vietnam, a hippie era, a culture of encouragement and defiance around drugs. And, adding to that, Greenfield's family was under duress: His dad, a graphic designer, and his mom, an art teacher and art therapist, were on the rocks. A very unstable marriage, with four kids, David being the oldest and most sensitive.

By fourteen, David was experimenting with drugs. Marijuana, barbiturates, LSD. Then he started taking and selling painkillers from his parents' medicine cabinet.

"It was a cry for help," he says, looking back. "My parents' marriage was crumbling."

He spent four and a half months at a place called Harold House, an inpatient rehab center in an old warehouse. He got clean. Still, he said, his high school guidance counselors had "thrown in the towel" on him.

They were wrong. David became a psychologist, earning a PhD in 1986 from Texas Tech University, and, eventually, becoming an assistant clinical professor of psychiatry at the University of Connecticut School of Medicine. He now brings a particularly personal perspective to the debate about addiction. He's the director of the Center for Internet and Technology Addiction, one of the first places in the world to treat technology addiction as a medical disorder.

What's happening today with technology, he argues, is comparable to what happened in the seventies with drugs. "It's exactly the same, the pace of adoption of technology and cultural acceptance isn't that much different than the pace of adoption and cultural acceptance of the drug culture, except that one is legal and one isn't."

HE GETS RAISED EYEBROWS. At conferences, people say: Gimme a break; technology's not like cocaine or heroin or crack, where you can see the tissue damage in the brain. They say: There's no proof of increasing tolerance levels to technology. And: People can walk away from their devices.

Dr. Atchley, while an admirer of Dr. Greenfield, isn't convinced that technology is addictive, per se, though Dr. Atchley remains open to the debate.

The so-called bible on the question of psychological illnesses is the *Diagnostic and Statistical Manual of Mental Disorders*, the *DSM*. The disorders, and their classifications, change with each edition, reflecting new science and understanding. Its committees have grappled with the question of technology addiction but haven't made it an official diagnosis. For now, it falls into a broader category, "Impulse Control Disorders Not Otherwise Classified."

According to a 2012 article titled "Are Internet and Video-game-playing Addictive Behaviors?," scholars from Yale and University College London concluded that the features of an impulse control disorder are "a failure to resist an impulse, drive or temptation to perform an act that is harmful to the person or others." It goes on: "The individual feels an increasing sense of tension or arousal before committing the act and then experiences pleasure, gratification, or relief at the time of committing the act."

To an extent this is a semantic conversation. The Yale paper notes that the word *addiction* derives from the Latin *addicere*, which translates to "enslaved to" or "bound by." They are broad definitions. It was only in the last thirty or so years that the definition came to be narrowed to substance abuse, notes Marc Potenza, a psychologist at the Yale School of Medicine and an expert in addiction science who was one of the paper's co-authors.

In short, most researchers don't put texting, or video game playing or Internet use in the same category as addictive drugs. They might be compelling, say many researchers, but not addictive.

To Dr. Greenfield, it's mere semantics. "Whether the word is 'impulse,' or 'compulsion,' or 'addiction,' clearly there is an overtaking of rational, logical processing of information and judgment like we see with other drugs," he says.

He's worth hearing out. For one, the way he describes and breaks down our day-to-day interaction with our devices is extremely resonant. His analysis rings true. Also, even though technology is not classified as "addictive," some neuroscience points to stark similarities between how technology use and drug use trigger chemical release in the brain.

IN 1998, RIGHT ABOUT the time that Dr. Greenfield sensed the lure of computer technology, neuroscientists at the Imperial College School of Medicine in London observed the brains of eight male subjects playing a video game.

The game entailed using a computer mouse to direct a virtual tank through a battlefield. Subjects had to collect flags while avoiding and destroying enemy tanks. As subjects got more flags, they progressed to a new game level. And they got a reward of seven pounds for each new level achieved.

The players were injected with low levels of a chemical called raclopride. The significance of raclopride is that when it travels through the bloodstream into the brain—crossing the "blood-brain barrier"—it attaches to dopamine. The chemical is also a radioactive isotope, which means it can be visualized using a PET scan.

The technique allows researchers to take a picture of the inside of a body, sort of like an X-ray. But one key difference is that a PET scan lets researchers look at various neurotransmitters, and cellular activity—a power unimaginable just a few generations ago.

The results were interesting, if open-ended. Dopamine levels at least doubled. And not just that, the subjects who performed better at the game had greater increases in dopamine.

To Dr. Greenfield, the true believer, it is a crucial piece of baseline evidence. "When you're playing a computer video game, the dopaminergic centers light up like a fucking Christmas tree."

DOPAMINE CENTERS ARE CRITICAL. They are, in a basic sense, our reward centers. They help tell us when we've done a good thing. They light up when we eat, or have sex, when we accomplish something. They are part of what helps us survive.

But they also light up when the brain interprets something as pleasurable, even if the behavior doesn't appear to have the survival value of, say, eating or procreating. They are activated in lots of circumstances, even when we're doing something that can be destructive. For instance, the reward centers light up when someone takes drugs, like cocaine, or booze, depending on a person's susceptibility.

Different drugs use different mechanisms to trigger increased levels of dopamine. Some addictive drugs, like cocaine, appear to prevent dopamine from being absorbed. That leaves more of the neurochemical in the synapses. Other drugs, like amphetamines, appear to induce a greater initial release of dopamine.

To Dr. Greenfield, technology behaves more like the amphetamine model. The way he thinks about it is that even the smallest click of a device gives a little rush, a tiny dopamine squirt. Hit a key and something happens. You click and a letter appears on the screen, for instance, or a picture comes up, an email opens. Each of these, on some level, triggers tiny rewards.

"It's in a sense a narcotic."

After a while, he says, the mere presence of the device begins to offer the promise of tiny hits, and bigger ones. It's what Dr. Greenfield calls "the anticipatory link."

It's a bit the way a smoker feels a little thrill when opening the pack, or lighting up, knowing the nicotine hit is coming. "You see the computer, it's one trigger, then you sit at the keyboard, it's another, you push the key, you get a result, then you get the big result. There's a cascade of dopamine. It's the big kahuna."

This is, in essence, interactivity. Touch the key, get a response; touch the screen, get a burst of information, or a reward. That's not inherently bad. But the way Dr. Greenfield sees it, people feed the need, click after click. And then, when the rush of excitement fades, he says, people feel rotten. "So they go try to get more."

A DIFFERENT STUDY, REPORTED in a 2012 paper in the *Journal of Biomedicine and Biotechnology*, found a connection between Internet use (rather than video game playing) and dopamine.

The researchers used PET scans to examine the brains of five men, around the age of twenty, seeking treatment for Internet addiction (use of the Internet more than eight hours per day) from the Peking University Shenzhen Hospital. The men experienced reduction in the dopamine transporter, a protein, as compared to a control group of nine men who were not compulsive Internet users. What's significant about the dopamine transporter, among other things, is that it also gets altered in people with chronic substance abuse. In other words: Some of the same pathways and results seen in the abuse of substances is happening in the brain of heavy Internet users.

The paper concluded: "These findings suggest that [Internet addiction] is associated with dysfunctions in the dopaminergic brain systems and are consistent with previous reports in various types of addictions either with or without substances."

It is worth noting the significance of the phrasing *"various types of addictions either with or without substances."* That word choice is crucial because, while there is interesting emerging science around Internet addiction, there is a predicate question being asked by many scientists: Can behaviors be addictive, or just substances? (At the center of the debate is a behavior like gambling; is it classically addictive, or does it belong in some other category of compulsion?)

Researchers point to a number of surveys of Internet users that suggest, at least, that their behavior can fairly be defined as "addictive." The 2012 paper from Yale ("Are Internet Use and Video-game-playing Addictive Behaviors?") summarized the findings of surveys and questionnaires. The surveys found a range of prevalence of "Internet addiction" among young people and adults across the globe. For instance, a 2011 survey of students in the United States found a 4 percent prevalence; a 2008 survey of elementary and high school students in Hong Kong found a 19.1 percent prevalence; others showed 10.7 percent among students in South Korea and 18.3 percent among college students in the United Kingdom.

There was a similar range in surveys exploring excessive video game use, or problematic video gaming (PVG).

The paper also found interesting evidence that people who identify as Internet addicts also tend to have personality traits, or psychological conditions consistent with substance abuse and "pathological gambling." These so-called comorbidities—meaning the conditions co-occur with Internet addiction—include "attention-deficit hyperactivity, mood, anxiety and personality disorders."

The paper says that, as with substance abusers, studies of people with Internet addiction have found "increased novelty seeking, low reward dependence, impulsivity, high risk taking, low self-esteem and disadvantageous decision making."

What that implies, but does not prove, is that some individuals could be more susceptible to Internet addiction, just as some people are more susceptible to pathological gambling or substance abuse.

The paper also explores the neurological studies, including one that suggests the release of dopamine by players of a computer-based racing game is similar to that produced by amphetamines and crystal meth. The Yale paper strikingly concluded: "Taken together, these findings suggest that (Internet addiction) is associated with dopaminergic neural systems in a fashion similar to substance-related addictions."

WHEN IT COMES TO the lure of technology and the way it stimulates people, there's one more comparison that researchers widely cite. It's not to drugs but to gambling, specifically, to slot machines. And the parallel stems from a concept that is quite counterintuitive: the Internet, smartphones, and other devices are addictive because they often deliver us worthless information.

Say what?

An American author named Frank Scoblete once wrote that slot machines "sit there like young courtesans, promising pleasures undreamed of, your deepest desires fulfilled, all lusts satiated."

In fact, what makes slot machines so powerful, at least in part, is that they so often leave the player unsatisfied. And, even more to the point: The players never know when they will get a payoff, a feeling of satisfaction, a fulfilled desire. In a nutshell, slot machines work on a principle called variable or intermittent reinforcement. Take, for instance, several classic studies with animal models. A baboon, say, is shown that if it pushes a lever then some food will drop through a dispenser. But the animal doesn't know which push of the lever will be the one that will deliver the food.

"The baboon will press the lever at a very steady rate. 'Is the food there yet, is the food there yet?' Each press is like a question," explains Dan Bernstein, a professor of psychology at the University of Kansas, where he has an office down the hall from Dr. Atchley.

It may not be a comfortable comparison for some. But the image of a baboon pulling a lever for food is not all that dissimilar from a person obsessively pecking at their phone waiting for the next email to appear.

And it can almost be assumed that much of the stuff that comes into our devices is not particularly useful. It is, in a word, spam. A 2012 report from Symantec, a company that builds software to block computer viruses, found that around 67 percent of email is spam. The big number probably comes as no surprise and, doubtless, much of those unwanted missives are blocked. But even if only a fraction get through to the end user, it puts a fine point on just how hard it is to know what you're going to get when the computer—or phone—pings with incoming information. Classic variable reinforcement.

And, setting aside the question of raw spam, the plain fact is that some information is simply more interesting than others. Some texts, calls, email, and Facebook status updates are really informative or entertaining. But you can't really discern who or what quality information is coming your way without diverting attention. Every time a text comes in, "You don't know what it's going to be, who it's from, and, hence, how valuable it is," says Dr. Greenfield. "What's happening, in essence, is that you're constantly scanning your texts and email because every once in a while you are going to get a good one and you can't predict when that is.

"The Internet is replete with novelty and variability," he contends. "That's why Facebook is so popular. It's the fact that it's dynamic, and novel, and constantly changing."

CHAPTER 25

REGGIE

IN MID-JUNE 2007, REGGIE was at the Missionary Training Center in Provo. It was a place filled with excitement and some nerves; young adults amassed, in shirts and ties, taking in long hours in the classroom, preparing for a two-year voyage of maturity and zeal, the culmination for many of a lifelong dream, but also two years away from family. No cell phones. No visits home, no TV or radio. The focal point was the gospel. Full stop.

Everyone got a name tag, on which the most prominent words were CHURCH OF JESUS CHRIST OF LATTER-DAY SAINTS. Then there was your own name, slightly smaller, denoting that this was not about the individual, but the larger quest.

ELDER SHAW.

Reggie was thrilled when he put it on. Then more good news. The district president named him district leader, in charge of leading ten other kids in activities. He wasn't sure why but assumed it was because he'd been to the center before, albeit briefly, and because he was a little older. On the very first day, he lead his group into the cafeteria for lunch and sagely directed them to the shortest line.

He said his group members just loved it. They had no idea he'd been there before. They were impressed. "I remember knowing the kids in my district loved being with me."

At night, it was four to a room, sharing two bunk beds. Reggie was sleeping great. Peaceful.

A week in, he was sitting in class when there was a knock on the door just before lunch. A man who helped run the program poked in his head. "I need to speak with Elder Shaw, please."

Reggie felt a surge of panic, "a pit in my stomach."

In the hallway, the man told Reggie there was a phone call for him. "It's a lawyer named Mr. Bunderson."

They took an agonizing walk to the main office. Reggie well knew he wasn't supposed to use the phone, and that it must be important if he was being asked to take a call, and, no less, from his lawyer.

"Hello," Reggie said.

"Hi, Reggie," Bunderson began, and immediately headed off the terror. "Everything is still going well."

Reggie just listened.

"I'm calling because I need you to sign some papers. They're called power-of-attorney."

Bunderson explained that he needed the papers "just in case" he would need to deal with something while Reggie was away on his upcoming mission. "If we did this, we won't need to call you every time I need something," Bunderson said.

Reggie felt more than a sense of relief. It was as if Bunderson were saying: *Go on, nothing's really happening, and I don't much expect it to.* That's what Reggie heard, at least.

"After the conversation, that's when I felt safe."

TONY BAIRD WAS STRUGGLING. Finally, he was in a position to truly confront the issues in the Shaw accident. Yes, of course, the accident was a tragedy. But accidents happen. What were the facts? What was the law? And what was the responsibility of a responsible prosecutor?

It was not uncommon for Baird to answer someone who asked about his job, "By the stroke of my pen, I can ruin a life.

"Just because you can do something as a prosecutor, doesn't mean you should do something," he says. "Plenty of things go wrong in society. There are plenty of things that are troubling to society, but you have to take your power seriously."

As Singleton began to amass evidence, and Baird saw that a potential case was coalescing, he let these thoughts percolate. He had a good familiarity with the driving laws, the reckless driving laws, the negligent homicide laws. And in the previous few years, he'd had cause to make hard decisions about how to use his prosecutorial powers in difficult cases. In one case, about six years earlier, Baird had prosecuted a twenty-one-year-old man who had been driving a small Toyota pickup through Logan Canyon, along a narrow, sometimes dangerous road. Earlier in the day, the man had told friends that his brakes were bad, Baird said. He was driving eighty miles an hour when he drove head-on into a family, killing the mother and grandmother.

The driver's grandfather had recently retired as a district court judge, which gave the accident an extra public layer. In the jury trial, Baird argued that the defendant drove recklessly even after knowing his brakes were bad. The conduct, no question in Baird's mind, rose to the level of "criminal negligence." In Utah, as generally elsewhere, the meaning of that standard is that the defendant shows a "gross deviation" from the standard of care.

The person *should* have known that their conduct was wrong and didn't just deviate from what was right, but substantially deviated. Acted in veritable defiance of law and common sense. The standard says that, whether or not you are aware of the risks, "you should have been aware," Baird explained. (By contrast, a greater standard, and tougher charge, is "reckless" behavior. Here, a person actually is aware and disregards the proper behavior.)

In the case of the driver with the bad brakes, Baird succeeded in securing a conviction of two counts of negligent homicide. The charge is a misdemeanor, but a serious one. The driver got ninety days in jail and a black mark for the rest of his life, in addition to the weight of his deadly actions.

Then, four years later, Baird prosecuted another trial that provided an insight into Baird's thinking as he considered the Shaw case. The second case concerned a man in his twenties who was driving a truck with his wife in the passenger seat down Blacksmith Fork Canyon, a steep, windy road. The night before, they'd been camping with friends, had stayed up late, the evidence showed, and were driving back in a hurry to get to work. The driver lost control, and drove off the side of the road, slamming into a tree before coming to rest in a river. The man survived. The wife's head was crushed by the tree, and she died.

On its face, it was an accident, but there was a twist, one that really bugged Baird. The evidence showed that the driver had smoked marijuana the night before, even though the toxicology report didn't prove that the man was under the influence.

Still, Baird was piqued. "He was willing to jump in the car, with supposedly the person you love more than anyone on earth, and drive too fast after a night of partying," he says, looking back. The case, on some level, was one that involved character. *What sort of person would behave this way?* Baird thought.

And the case involved some cultural issues, the frowning upon of drug use, the lack of appreciation, thought Baird, for how it can impact the brain and behavior, even after it's mostly out of your system.

Baird conceded, though, that because of the toxicology issues, this one may be tougher than the one with the driver and the bad brakes. Baird brought the charges anyway and, again, he won; another negligent homicide conviction, and a sixty-day sentence.

Then there was Reggie.

First of all, Baird thought, there was just the newness of the whole concept of texting and talking on the phone, and its implications for driving. As Baird found time to explore the issue, he came to the conclusion: "There was no other case law, and nothing statutory that gave us any real guidance as to whether this constituted negligent homicide or something else."

There weren't great stats on how many wrecks were caused by cell phone–using drivers. There were mounting anecdotes. Just a few days earlier, across the mountains in Idaho, an eighteen-year-old driver had been sending a text message when she lost control of her car, jettisoning a fifteen-year-old passenger through the windshield. The girl was in the hospital in serious condition.

Baird was thinking about the question of whether Reggie—presuming he was, in fact texting—should have known his behavior was reckless. After all, Baird thought, if the technology was new, was it fair to make the case that Reggie was grossly deviating from a standard of care, if the standard of care wasn't clear?

Was he really on notice? Baird asked himself. He said he himself knew the conduct was dangerous and wrong, but did Reggie? Should he have?

"Texting is not a dinosaur like vehicles and driving. It is such a new phenomenon," he said. He was wrestling with other questions as well: How is texting different from looking down to unwrap a piece of gum, or to sip from a Big Gulp? These are things that, tragically, lead to a momentary loss of focus; but they are accidents, tiny blips, and instants.

Eating, smoking, changing the radio.

Or how about waving? There was something in Baird's past, when he was a young man. He was on a motorcycle, and he waved to a mailman and . . . he didn't like to think about it. It was the worst tragedy of Baird's life. People made mistakes, right? Still, it wasn't time to think about that, what he'd been through, or its implications on a potential Reggie Shaw trial.

This was about the facts, and the case, and the law. His initial impression remained: "There was no precedent."

"MAY I GIVE YOU some information?" Reggie asked the woman.

She said yes. He gave her a pamphlet about the Church of Latter-day Saints, and a phone number to call if she wanted to follow up.

Reggie felt euphoric. His first pitch seemed to have gone well. And he hadn't even landed in Canada to officially start his mission. He was still on the airplane, a Delta flight, from Salt Lake to Winnepeg.

At the Salt Lake airport, there had been back pats and "good lucks" from other travelers who recognized the band of suit-clad missionaries. On the plane, Reggie sat in an aisle seat, and struck up the conversation with a woman who asked about where he was going, and why.

"It was the first time I actually attempted to teach someone who I didn't know, who was not a member and not familiar with the Church," Reggie said. "She listened well, she asked a lot of questions."

The last thing she said was: "I'll take this information to my pastor and we'll talk about it."

Reggie recalls her reference to her pastor with a laugh. "That's never a good sign."

He was feeling buoyant, on his way, and the pitch to the woman allowed him to think it might involve fewer doors shut in his face than he'd been prepared to expect.

In Winnepeg, he met his "companion," the guy he'd be paired up with, walking the streets, suit-clad, scripture in hand. He was Elder Smith. He was on the tail end of his own mission and would be showing Reggie the ropes. They hopped in Smith's small Ford and took the drive to Regina, a big city that would be their home base.

Just a day in, Reggie was walking down the street with Elder Smith, going house to house, when a car drove by. Someone yelled at the pair, and then a bottle came flying in their direction. It missed. It shattered. It was a bottle of SoBe, an energy drink.

Though he'd come to expect some of this, he was kind of shocked. To Elder Smith, it was just another day.

He thought: Any time you're trying to do good, Satan is going to try to stop you. It might be with a SoBe bottle. Anything to frustrate or derail your path.

CHAPTER 26

TERRYL

NINE MONTHS EARLIER, TERRYL had approached Jackie Furfaro outside gymnastics practice to ask if she needed anything. Jackie had said no.

Since that time, Jackie and Leila each had kept in periodic contact with the investigators, expecting little, getting even less. Leila was in more regular contact, speaking periodically with Singleton. In June, in one of those conversations, she got the impression from him that he was frustrated and that it seemed the likely outcome of the case would be a ticket for driving left-of-center. She called Jackie, with whom she'd had little interaction.

She explained her frustration, and Jackie shared it. Jackie decided to call the prosecutor's office. It was late June. Terryl wasn't around. She was outside Washington, D.C., packed into two rooms at the Days Inn with the whole family and her mother for a very special event: Jayme was the Utah state representative for National History Day—a competition in which more than five hundred thousand students compete. There was extra meaning in the event for Terryl because her real father, and his wife, joined the group, marking the first time the whole collection of them had been together.

Shortly after Terryl got back home, in early July, she got a call patched through from the receptionist while she was sitting in her office, in the basement of the Cache County prosecutor's office.

Terryl's basement office had windows, but you sort of had to crane to look out of them because the room was half underground. Three plants in pots graced the window facing east toward the front of the courthouse. Through the north-facing window, she could glimpse the Best Western that was kitty-corner to the courthouse. It was a far cry from the ornate offices upstairs.

"Hi, Terryl. It's Jackie Furfaro."

"Jackie!" Everything Terryl said seemed to have an exclamation point on the end.

"I could use your help."

Jackie explained about the accident. Terryl listened for fifteen minutes or so. She took a few notes. That was all she needed. This was right in her strike zone: an aggrieved victim, a remorseless alleged perpetrator, and no one taking up the fight.

She started researching in earnest. She popped in on Baird and his boss, George Daines, the county attorney. On July 6, nearly finishing her research, she made a handwritten note in a hybrid of script and cursive, letters smaller and neater and more right-leaning than in her diary. She noted a few things, like: "driver uncooperative," and "text messaging while driving (cause of accident)." She wrote down possible charges, including manslaughter, which is a felony, negligent homicide, and reckless driving. She asked a question: "How many X did John see (driver) go left of center," referring to Kaiserman, the farrier who was behind Reggie in the car.

She did more research. She wrote a memo.

THE JULY 6 MEMO was written in a style that her colleagues think of as quintessential Terryl: deeply passionate but somehow dispassionate enough, and just to the point. It was written to Baird and Daines.

"I know I have discussed this case with both of you," it began. The first paragraph went on to remind the lawyers that she'd been meeting with the victims' families. The paragraph concluded: "The victims' families are asking that Negligent Homicide Charges be filed."

Then Terryl listed twenty bullet points over two pages. Among the facts: According to phone records, Reggie texted Briana Bishop five times on his drive and received six text messages from her. In all bold, Terryl wrote:

Please be aware that rain is falling and the roads are slick and wet.

Terryl recounted in short bullet points what Rindlisbacher saw at the post-accident scene, including his witnessing of Reggie texting.

"Reggie *lies* to Trooper Rindlisbacher and denies using his cell phone for calls or texting while driving that morning," Terryl wrote. This is a supposition, albeit a reasonable one. The records show what Reggie was doing, and it was fair to assume he'd lied rather than forgotten what happened because of the trauma.

At this moment in the overall narrative of the accident, Terryl gives voice to a new kind of emotion: Indignation. Barely masked.

I think that the most disturbing thing to me is that even though he knows there is a criminal investigation taking place (his family has hired Jon Bunderson), he still goes ahead and submits his mission papers.

And she continues, after a few more sentences, that the sum of the misconduct of Reggie and his allies enabled Reggie to go on a mission, which could somehow protect him from justice.

Because he has lied to law enforcement . . . and he has hidden behind an attorney, he will not have any consequences for his actions as he is now on a mission.

And:

The bottom line here is two men are dead because Reggie felt that it was okay to use his cell phone for text messages and Reggie obviously thought that it was okay to lie to a trooper.

In a subsequent paragraph, Terryl observed that the Utah public schools have a class on distracted driving in their driver's education class. In all bold, she offered the following supposition:

I believe that the reason Reggie lied to Trooper Rindlisbacher and in his written statement is because he knew it was dangerous to be sending/receiving text messages while driving 55 mph in the rain with wet roads.

Then Terryl began to touch on the science to back up the dangers of texting and talking while driving. She cited Dr. David Strayer, the University of Utah scientist who'd failed to get the cell phone industry's attention and then struck out on his own. He'd, by now, devoted himself to showing the risks to attention from using a phone behind the wheel. His evidence was mounting and profound; this was something different, this device that captivated the human senses, and, in a way, recognition of the risks required neither science nor proverbial rocket science.

The research showed many people intuitively knew of the risks of talking on the phone but did it anyway.

Terryl summarized all this, and then added a final sentence. It was the kind of thing that just drove her nuts. Reggie, she wrote, "has gone on with his life and has not shown any remorse for the extensive damage he has done."

ON JULY 17, TERRYL invited Jackie and Leila to the offices to meet with Daines and Baird. It was early, seven a.m. They gathered in the conference room, a quietly intense and splendid setting. It's not so much the insides that demand attention—the room is rectangular, with a conference table surrounded by beige cloth–covered chairs, and with American and Utah State flags in the corner—as it is the view.

The window, facing east, provides a perfectly framed glimpse of the Logan temple, which stands less than half a mile away, and the majestic Rockies behind it. The temple was only the second built in Utah, constructed by volunteer labor over seven years starting in 1877. The exterior of rough-hewn limestone was originally painted white, according to the Church website, but the paint has been allowed to weather away, and now the building tends toward brown; at a distance, its body blends into the mountains.

But not its steeple. White, almost gleaming, and certainly ornate, the steeple seems to project from its top into the sky.

That morning, the sun rose over the mountains, peeking in through the conference room window. Jackie and Leila told the prosecutors how they felt. Jackie, according to Terryl's handwritten notes, said that Reggie, from what she'd heard, was just going on with his life, not taking responsibility, and it's "no big deal to him." Jackie said that Reggie lied about it. "How," Terryl wrote that Jackie wondered, "to teach my children about responsibility?"

For her part, Leila said she wanted to see Reggie punished, and to hear him say he was sorry. People can't text and drive, she said. "People need to know they can't do this," Terryl wrote that Leila said. This is a "preventable distraction."

The prosecutors took it in. They were nice but noncommittal.

A day later, July 18, an intern in the office—a second-year law student named Jacob Gordon who would later join the office as a prosecutor—produced a three-page memo for Baird at the prosecutor's request. The subject: "Reggie Shaw, Possible Charges."

At the top, it said, "The Issue: What is the appropriate charge for a defendant when he is texting on his cell phone and causes an accident that kills two individuals?"

The intern wrote that there were four possible charges: reckless driving, reckless endangerment, negligent homicide—all misdemeanors—and manslaughter, the one felony option. He defined each charge.

Reckless driving: "operating a vehicle in willful or wanton disregard for the safety of person or property."

Reckless endangerment: "under circumstances not amounting to a felony office, the person recklessly engages in conduct that creates a substantial risk of death or serious bodily injury of another person."

Negligent homicide: "the actor, acting with criminal negligence, causes the death of another." And he notes that in this definition "the [actor] *ought* to be aware of a substantial and unjustifiable risk that the circumstances exist or the result will occur. The risk must be of such a nature and degree that the failure to perceive it constitutes a gross deviation from the standard of care that an ordinary person would exercise in all circumstances as viewed from the actor's standpoint."

Manslaughter: "recklessly causing the death of another." And, the intern added, that recklessly here means "when he [the actor] *is* aware of but consciously disregards a substantial but unjustifiable risk that the circumstances exist or the result will occur. The risk must be of such a nature and degree that it constitutes a gross deviation from the standard of care that an ordinary person would exercise under all the circumstances as viewed from the actor's standpoint."

Clearly, Reggie's actions had taken a life. That wasn't in dispute. But did he know of the risks, and should he have known of the risks? The law says the penalty is harsher if someone should have known of the risks and harsher still—manslaughter—if the person does know the risks and wantonly disregards them.

"To be sure," the intern wrote, "I think the defendant is guilty of manslaughter." However, he wrote: Unless Reggie were to admit that he knew about the risks of texting while driving, he would be difficult to convict.

And so, Gordon, the intern, settled on negligent homicide, the toughest misdemeanor penalty, arguing that the two charges for reckless behavior are not "harsh enough to do justice to the outcome of the accident."

But there was a caveat. A big one that took up the third page of the memo. The caveat: There was no precedent. He wrote, "There is virtually no case law whatsoever in the entire United States that has addressed it. The technology is just too new."

Washington, he wrote, is the only state to have passed a law banning drivers from texting while driving, while a handful of states had banned such behavior with drivers with learner's permits. Four states banned talking on a handheld cell phone.

In his conclusion, he emphasized Reggie should be charged with negligent homicide and also reiterated: "There is no supporting case law in the United States at this time, owing, most likely, to the relative newness of the technology."

JUST A FEW DAYS earlier, across the country, there was another tragic example of the stakes. In New York, five high school cheerleaders died in a head-on wreck with a tractor trailer. Police suspected texting.

According to the police, as reported by the Associated Press, the driver of the car sent a text at 10:05:52. She received a reply at 10:06:29. Thirty-eight seconds later, someone called 911 to report the fiery, deadly wreck.

BAIRD WAS APPRECIATIVE OF the law student's memo, but concerned. This was not an easy case, with little or no precedent and a potential defendant in Reggie who may well not have known the risks of his behavior. Even if Reggie knew, or suspected, how would they prove that? To Baird's knowledge, there had been no big movement to warn people. Heck, Baird himself had barely heard of the concept of texting and driving. What would a jury think?

And Reggie was a nice kid, wasn't he? Just a regular guy, churchgoing, on a mission, clean-cut. He lied. But people lie, Baird thought, when they have their backs against the wall. He was like any of us, Baird allowed himself to think; any of us could make a mistake like this.

He needed more information. On July 23, he wrote a memo to Singleton, the investigator, asking him to detail what Kaiserman the farrier had seen.

That same day, Terryl was forging ahead. She wrote another memo, this one five pages, detailing the facts on the morning of the accident. She couldn't get the case out of her head, like so many others that had grabbed hold of her over the years. But this one really irked her, and it wasn't the texting and driving piece, at least not first and foremost.

"What bothered me about this case was that he didn't say he was sorry. It really bothered me. He didn't apologize and he didn't acknowledge what he had done. I know George [Daines] told me he had a good attorney who was working for him, but he didn't even acknowledge that he killed two men.

"When apologies are needed, people are owed apologies."

CHAPTER 27

THE NEUROSCIENTISTS

WHAT DOES ALL THE evidence about the powerful lure of personal communications technology add up to?

Researchers answer by way of an analogy. In the twenty-first century, technology can be compared to food. You need food to live. And while you may not require technology to survive in the same way as food, our cultural and social lives require it.

One of the great innovations in modern times has been the industrialization of food. No longer must people till their own gardens to eat, or hunt for their own meat. Someone else does that, and we buy it at the store. As a result, we have much more time in our day to do something else—to create, spend time with our families, play, build and improve societies.

But the downside of that industrialized food process is that, if we're not careful, we can succumb to the ease of access. Eat too much, and you get obese. Eat the wrong things, you get sick—everything from a stomachache to cancer to heart disease. But tempting us is fast food, or a vending machine that is only steps away, providing in a single bag of potato chips the fat and sugar content that, generations ago, we would've had to work all day to discover, prepare, and consume. In the old days, by the time we found (even fought for) the calories, we'd earned them. If we are not careful, the ease of access to high caloric (salty and fatty and tasty) food, will wind up using our most primitive cravings against us.

There is a pointed parallel to personal communications technology. In the same way we crave food, we crave connection. Not just for its own sake but because connection is essential for survival. It helps us form networks, understand sources of opportunity or threat, create alliances, fight enemies. It is primal. "We are a kind of animal that has evolved for social learning," says Nicholas A. Christakis, a physician and Yale professor, who is an expert in social networks across time. By way of example of the value of social connection, he notes the incredible value of, say, learning from someone that fire can burn, rather than each of us having to get burned to understand fire's danger.

Now come ultra-powerful devices that provide such easy communications that they can, if we're not careful, use our social survival skills against us. A simple example illustrates the point. There are few impulses as basic and inescapable as the one that urges you to turn around if someone taps on your shoulder. You must discover if the person is an opportunity or a threat.

When your phone rings, it is a proverbial tap on the shoulder. You want to find out who it is. You need to. Your bottom-up survival system demands it.

All the research by Dr. Atchley and many others has illuminated the idea that this new technology plays to our deepest human needs. When you add it all up—the social lure of information, receiving *and* disclosing; the intermittent delivery mechanism; the stimulation of interactivity and the neurochemicals associated with reward—you wind up with something powerful to the point of being overpowering. To some researchers, it feels like a process of neurological hijacking.

"The cell phone, and other similar technology, meets a deep need for social connection with a greater ease and greater potential detriment to it in the same way that a vending machine that is right down the hall plays to our need for calories," Dr. Christakis says.

He and others point out that digital communications are not alone throughout history in raising questions about how technological change impacts life and interaction. The printing press raised questions about whether we'd be overwhelmed with information; the telephone raised questions of whether people would lose face-to-face interaction. But with digital technology, researchers say the amount of information, the speed of its delivery, and, pointedly, its interactive nature, have changed our world by orders of magnitude.

"We use stone-age brains with space-age technology, and that can lead to trouble," said Daniel E. Lieberman, an evolutionary biologist at Harvard University. Our tech tools let us be "hyper social," he says, which has many benefits, and also costs. "We're using them with brains not geared for this sort of thing."

In addition to comparing digital technology to food, there is another compelling analogy. It is far-fetched in some ways, but also highly illustrative. It compares technology to the immune system, which is, of course, crucial for survival. It protects the body, helps defend it against intruders. But sometimes it can spin out of control and produce antibodies that attack the body itself. The survival mechanism has become the enemy, as in such illnesses as lupus and rheumatoid arthritis that can attack organs, joints, even the brain. So, too, our personal communications can, if out of control, turn a powerful survival tool into one that works against us. That's not to say they will kill us, or lead us to kill someone else, as in the case of distracted driving. But they may work at decided cross-purposes to what they are designed for. Rather than assisting in our stated goals—productivity, creativity, long-term thought—the tools can undermine those crucial aims.

The top-down goals become swamped, overwhelmed by the primitive bottom-up system aimed at warning us of opportunity or threat, playing to our deepest social needs. The thing is, the information that comes in may not be crucial at all; it may actually be irrelevant, spam, but, weirdly, that seems to reinforce the compulsion to answer the device. In fact, the mere pressing of buttons is rewarding, providing a dopamine release. This primitive warning system that is meant to serve us, can, instead, enslave.

"When the phone rings, it triggers a whole social reward network. And it triggers an orienting response that has been wired into us since hunter-gatherer times. You had to pay attention for survival. If you didn't attend you got eaten by lions. We're hardwired that way, no matter what we want to do. It's extremely difficult to turn those things off. It's in our DNA," says Dr. Strayer. "Engineers have co-opted these devices to have those very signals we can't ignore."

SO IF THE CALL of junk food leads to overeating and too much obesity, what does too much interaction with our device lead to? What are the downsides?

For one, researchers worry that heavy use of interactive media can, over time, reduce attention spans. The fear is that we grow so accustomed to frequent bursts of stimulation, we have trouble feeling satisfied in their absence. This effect could be true even if the bursts are not addictive, merely habit-forming. Think about it: You hear the ping of an incoming text or call, you respond; the ping happens, you respond. And each time you respond, you get a hit of dopamine. It's a pleasurable feeling, a release from the reward center. Then it's gone. There is no incoming text, no stimulation. You start to feel bored. You crave another hit.

Chasing a dopamine hit runs counter to focus and goal-setting; needless to say, it becomes hard to sustain periods of attention.

This is something that seems particularly true of young people, children, and adolescents. This concern gets wide agreement among researchers, even when they disagree about whether our gadgets are actually addictive, or merely compelling. Dr. Potenza, the Yale scholar who has a more cautious approach than Dr. Greenfield or Dr. Atchley, says, "We've led to a generation that is perhaps less tolerant of waiting for delays. It's hard for them to occupy themselves without the same degree of stimulation."

Dr. Greenfield, predictably, goes further. He deems young people who are raised on digital devices "Generation D." "They're so amped up on dopamine that when it's not firing, they feel dull, dead," he says. And that means they need to move on to the next thing, quickly, rather than staying with something. "They have no threshold for attentional capacity."

The concern isn't limited to children, but many scholars say they are more vulnerable because their frontal lobes are still developing. What that means is that even in the absence of interruption, they are still developing the capacity to set and sustain goals. Add in a device that rewards interruption and the challenge gets that much more intense, according to Dr. Michael Rich, executive director of the Center on Media and Child Health and an associate professor at Harvard Medical School. In an article in the *New York Times* about the impact of technology on young people's attention, Dr. Rich said: "Their brains are rewarded not for staying on task but for jumping to the next thing."

While the broader concerns about attention span are widespread among scholars, the evidence can be indirect, some of it anecdotal. For instance, at the end of 2012, two surveys from highly reputable, nonprofit research groups—the Pew Research Center and Common Sense Media—each reported that American teachers believe heavy technology use is hampering students' attention spans and their ability to persevere through tough challenges. The Pew report, done in cooperation with several other nonprofit groups, found that nearly 90 percent of teachers said the Internet is creating "an easily distracted generation with short attention spans."

Another study suggesting falling attention spans found that we are even spending less time with the apps on our mobile devices. The study, by a research firm called Localytics, looked at the use of five hundred news apps between July of 2012 and July of 2013. It found that people were spending 26 percent less time in each session. At the same time, it found that people were launching their news apps 39 percent more each month.

It's important not to read too much into one study. But the research points to a very interesting dynamic: The reason people are spending less time in each app session isn't because their interest in news is declining.

After all, people are launching news apps more and more often. They like and want news. They just don't have the patience to spend much time with it.

THERE IS ANOTHER INSIDIOUS way in which the power of our devices to capture attention could be impacting daily life. It could be compromising our ability to make decisions.

This concern has to do with the toll taken on our brains when they are overloaded with information. Going back to World War II, researchers had shown that we make more unconscious mistakes when we are overloaded (like pulling the wrong lever in an airplane). But newer research suggests that such overload can make it harder for us to make good decisions when we are confronted with clear choices. Take, for instance, a 1999 study that has to do with chocolate cake.

In the study, undergraduate students were given the task of choosing what snack they wanted to eat, whether a delicious yet calorie-rich chocolate cake with cherry topping or a serving of healthy fruit salad. The question was complicated by the fact that the students were asked to memorize certain information before they made the decision. Some students had to remember a seven-digit number and others a two-digit number.

What the researchers found was that students were more likely to choose the chocolate cake if they were asked to memorize the seven-digit number. As the researchers put it, "Choice of chocolate cake was higher when processing resources were constrained."

There is other supporting research that shows how learning, memory, and decision making get impacted by an overloaded brain. The frontal lobe, the executive functions, get so overloaded, so taxed, that there are fewer resources left to make a good decision. Put another way, says Dr. Atchley, the frontal lobe is crucial in helping people inhibit impulses, like choosing the chocolate cake, which is a very basic form of decision making.

Or, for instance, deciding whether or not to focus on the road or the phone when you're driving. If your brain is taxed, you may not even be able to make a clearheaded decision about what is the right thing to do.

"To make a choice, you need frontal lobes active and you need enough competitors in other parts of the brain so that you can engage systems to make a decision."

Add it all up, says Dr. Greenfield, and you get a picture of a teenager, with an immature frontal lobe, brain possibly too overloaded with information to make a good decision or just unable to resist the chime of the phone in the center console. For such a driver, says Dr. Greenfield, the act of picking up the device becomes primitive, bypassing higher-level thinking.

"His brain is flooded with anticipatory dopamine. He knows on a primitive, neurochemical level that he's about to get a squirt," says Dr. Greenfield. "That's why he pushes that fucking button. He's not conscious of it."

YEARS EARLIER, DR. TREISMAN spoke of the idea that our very sense of reality gets established by what we pay attention to. This is maybe the most far-reaching implication of the powerful lure of our technology: If it co-opts our attention, it could reshape our sense of reality.

It's not nearly as far-fetched as it might sound. It's based on a simple proposition: Our day-to-day reality is based on what we see and hear with what we experience. The thing about our electronic gadgets is that they can easily redirect our focus. They cause us to look down, switch what we're listening to, change what we're thinking about. When that happens, it changes our reality. Think of it this way: If a tree falls in the woods, but you miss it because you're lost in a video game on the phone, did the tree fall?

Or consider a more pointed example. Say you're driving down the road and your phone buzzes with a text. You look down to read it. You aren't looking at the road and slip across the yellow divider. You hit another car. But by the time you look up again, and realize what's happened, the moment of impact has passed. Did you cross the yellow divider? Did the other car cross over? Did you hit a slice of ice?

"If you read car accident reports, it's very common for crash reports to read: 'I was driving. I was paying attention, the person appeared out of nowhere. I was looking at the road, and the other car was there—like magic,'" says Dr. Atchley.

"There are two options—either people are lying to protect themselves, or experientially that's what happened. It's like someone's inattention literally has played a magic trick on them—which is that a car appears out of nowhere.

"The eyes are open but the brain's not processing all the information."

"CAN I BORROW YOUR phone?" Dr. Atchley says.

He's at the counter of the Hereford House steakhouse. It's where he's giving a lunchtime speech to the Engineers Club of Kansas City, just an hour after Maggie, the undergraduate student, tried to navigate the virtual driving machine while getting texts from Dr. Atchley's graduate assistant.

It's crisp and sunny in this suburb of Zona Rosa. The steakhouse, beige with red awnings, is in an outdoor mall with all the regular shops: American Eagle, Kay Jewelers, and the like. It's a chain restaurant, serving food that has the trappings of being fancy. The open-faced prime rib sandwich ($13.95) comes with au jus gravy. Dr. Atchley feels mildly nervous. He should: He's in the wrong steakhouse.

He was supposed to go to Zona Rosa but in a different location. "These were the Google directions," he mutters.

Maybe, or perhaps, he concedes, he just mixed up the different locations, given everything on his mind. In any case, now he needs to borrow a phone so he can tell the people waiting for him to speak that he's going to be late. And he needs to get directions to the other restaurant.

The woman at the counter offers her landline. But she's having trouble coming up with quick directions to the other location, in Leawood.

Dr. Atchley, clearly anxious, borrows a cell phone and taps the right location into the mobile Google search engine. It looks to be about twenty minutes away.

Reflecting on the value of the cell phone, he says, "It was a faster way to get the information." But, he adds, "there were other options."

As Dr. Atchley tries to find balance with technology and prevent it from overtaking him, he finds there are occasions when his boundaries don't serve him. Recently, during a heavy snowfall, he bought a second phone so he and his wife could each have one if they needed to communicate or got stuck in a jam, rather than continuing to share one phone.

"We wanted her to have some way to call emergency services if she got stuck," he says. "Why not have that device?"

It seems like a silly question, rhetorical, for most people. Why not have the convenience? And Dr. Atchley really feels that way, too, despite the fear of temptation.

"I'm not a Luddite," he says. He points out all the technology in his world: the Wi-Fi connection at the underground house, the machines he uses to study the impact of technology—the brain scanners and driving simulator and computers he uses for everything from writing papers and doing research to complex tasks like doing statistical analysis. He emails with students and shares research with colleagues online. He can get lost for hours on Reddit, a tech-centric website he loves.

This new generation of neuroscientists isn't antitechnology, despite their cautionary findings. Not Dr. Strayer, certainly not Dr. Gazzaley, or any others of his ilk. They are tied to networks, reliant on phones, eager to communicate their messages and science, use technology to develop and further it. And then there are just everyday tasks; Dr. Atchley got lost heading to his lunch talk and a phone came in handy in finding the correct steakhouse. Once he got there, he plugged in his PowerPoint presentation so the audience could ooh and aah.

This is, quite obviously, not a question of either/or—do we either live with technology or give it up entirely?

"The question," says Dr. Atchley, "is how do we balance this stuff?"

CHAPTER 28

HUNT FOR JUSTICE

TERRYL'S JULY 6 MEMO, and her ongoing persistence were causing a stir.

It was time for the roundtable meeting with all the lawyers in the office of the Cache County Attorney's Office. This meant Terryl, Baird, and Daines, the county attorney. In preparation, Baird distributed copies of three statutes: the definitions of manslaughter, reckless endangerment, and negligent homicide.

Almost without discussion, the group tossed out manslaughter. They began to discuss the other options. It was clear from Terryl's memo where she stood.

Baird was convinced otherwise. In the meeting, he made a few points: This technology is new, and it's not fair to assume Reggie should've known better, at least in a way that satisfies the standard of criminal negligence.

"My mind-set was: Let's send a message, but do it with a different charge, a lesser charge," he says. "Let's send a message with this one— and next time we won't be so kind."

There was another factor in Baird's mind: Reggie "seemed like a decent kid."

By now, it seemed fairly clear that Reggie had lied—he must've known about the texts. But Baird wasn't that troubled by the apparent deception.

"As a prosecutor, people lie to me every day," he says. "When people get backed into a corner, they will say things that are not always the truth. They cast themselves in false light. I understood this kid was lying to us, but I really wasn't going to hold that against him."

Baird considered it a positive that Reggie was serving on a mission. Baird had done a mission himself, to the Philippines. It wasn't about the fact he and Reggie shared a common faith, he said. It was more the commitment to service; he says he'd look with equal favor upon someone in the peace corps.

"Maybe we could do reckless driving," he suggested. And he thought: "Negligent homicide carries such a stigma. I'm looking at this nineteen-year-old kid, thinking: This is a big burden to hang around his neck."

Besides, Baird had some empathy. Many years earlier he'd been a young man behind the wheel, momentarily distracted, with a life hanging in the balance.

BAIRD WAS SIXTEEN. IT was August. Farm country. Wide-open fields, dirt roads. Baird had recently gotten a Honda Nighthawk 650. He climbed on the motorcycle to go get a haircut, when his dad poked his head outside.

"Son, you better get your damn helmet."

It struck Baird as odd, his dad reminding him to wear a helmet. "We were not helmet people." This was partly familial and partly cultural in the sense that Utah is a place in which many people believe strongly in the concept of individual rights.

Baird drove down the long dirt road, tall grass on either side. He was going forty miles an hour. It was hot. He wore short waterskiing shorts and a white tank top. Not far into the ride, he saw the local mailman in front of a house. Baird waved.

"I turned my attention to in front of me," Baird recalls.

He saw the kid, the boy right there, right in the middle of the road. Baird could see the boy's face. The boy just froze.

Pop! Pop!

The boy's head hit squarely in the front headlamp.

Baird went somersaulting. Rolled a couple of times. He had so much momentum, he came up and was running. Then he heard the boy's mother coming out into the road.

"She picked him up and she just started wailing," Baird says. When he tells the story, he has to pause, for maybe a minute, choking back grief. He whispers, "Fudge," a kind of expletive.

The Utah Highway Patrol came to his house shortly thereafter to interview him, but mostly to reassure him. "The trooper sat down with me and my dad and he said: 'I'm sorry, son, there was nothing you could do.'

"It wasn't like I looked away for a long time," Baird states. And, besides, the boy just bolted into the road, apparently excited to retrieve the mail. "There was nothing I could do."

Baird says he thought about his case and Reggie's.

"I know what it's like to be a young kid who gets into an accident," he says. But he thought the similarities only went so far. "I don't know that you could compare the cases."

TERRYL WRESTLED WITH TWO thoughts: (1) She didn't agree with Tony, and (2) she respected him as much as any prosecutor she'd ever met. She'd come to the second realization some years earlier when they'd had a difficult back-and-forth over a tough case. Baird hadn't brought the maximum charge, and Terryl was upset. In the end, Baird had to remind her that her aim should be justice, not the maximum penalty.

It really moved Terryl, and, since that time, she'd tried to embody his message. *He's my favorite prosecutor,* she thought to herself.

But that didn't mean she always agreed with him.

The ultimate decision on how to proceed fell to Daines, the county attorney, though in practice this process was more democratic than tyrannical. As the conversation wore on, it seemed like he was siding with Baird. There just wasn't precedent here, and there were just so many resources in the office.

He drew out a middle ground.

"Terryl, if you can find a prosecutor to take it, then we'll pursue it."

In other words, he was leaving Terryl an opening.

"He knew what he was doing. He knew what would happen if he gave me an opportunity. He knew exactly what I was going to do."

THERE WERE SEVEN LAWYERS who worked for the county, not including Daines. He was elected, the rest were county employees whose job status was not reliant on elections. Among the seven others, Don Linton, the chief deputy county attorney, held relative primacy.

He'd been around for many years, working regularly with Terryl. He had a nickname for her—the "Sparkplug."

"The other attorneys knock. Everybody else knocks," Linton says, in contrast to Terryl's habit of bursting into her colleagues' offices unannounced. "She never knocks."

Terryl saw Linton as someone who didn't flinch in taking on unusual or tough cases; he'd had a particular interest in going after rapists and child abusers. Terryl figured Linton was her best shot.

Just a few minutes after the meeting with Baird and Daines, Terryl found Linton in a corner office opposite Daines', facedown in some papers.

"Terryl throws open the door and says: 'Hey, I need to talk to you.'"

"Okay," Linton said, both curious and bemused.

Terryl held a rose-colored piece of paper, which she extended to Linton. Without looking closely, Linton could see bullet points, an explanation of some kind.

"Don," she started. "I've got something you have to do."

Linton gestured to one of the armchairs—*Slow down, Terryl*—and she took a seat.

"These two rocket scientists were killed," Terryl began, as she launched into the story of the accident.

Up until then, Linton hadn't heard a word about Reggie Shaw, hadn't read about the case in the paper. He started to take it in, listening to the narrative and analyzing it from several sides.

From a legal perspective, his primary stance, he wondered, *What was the law here?* He couldn't and didn't make a snap judgment.

Instinctively, Linton also reacted to the story as a father. He and his wife had four children, two boys and two girls, the youngest being Libby, nineteen years old at the time. And Linton thought of her as a "terrible driver." A great musician, very bright, but not a good driver, Linton believed that she sometimes tailgated, and sometimes sped. And he knew that she texted, a lot; he suspected she did so when she drove.

He almost immediately started thinking about the case not just from the victims' perspectives but, as he puts it, from the "perpetrator's perspective." And also from the position of other possible perpetrators, including kids like Libby. "I was thinking about my daughter and her life and how it would change" if something like this happened to her.

Linton glanced at the memo Terryl had handed him and could see now that it was a set of facts and assertions mirroring what Terryl was telling him—an outline of the case.

As Terryl concluded, she said: "George isn't interested," referring to Daines.

Terryl finished by saying: "I'll do all the work, but you have to do this for me." Of course, that wasn't realistic, the victim's advocate couldn't do all the work.

Linton didn't commit. He told Terryl he'd do a little research. But she had piqued his interest, in a way he says likely wouldn't have happened, absent her intervention. In fact, he recalls, it was this conversation that set fate on its course. "When she came into my office," Linton says, "that was the spark that ignited the whole thing."

It is a sentiment echoed by Singleton, who says his efforts and conviction fell on deaf ears with the prosecutors until Terryl started pushing. "If it hadn't been for her, no one would ever have heard of this," he says. "It would've been a few months out of my life and a stack of paperwork that would've gone into the shredder.

"When you meet her, she's a very nice woman. She doesn't seem like a hard-core person," he says. "But if you want someone to go to bat for you, she's the one."

She was just getting started, and now so was Linton.

LINTON, LIKE THE OTHERS in his office, didn't know much about the subject of texting and driving. He considered the concept "a mysterious bog." He was fifty-two years old, a different generation, the last child of eight, who spent his first twelve years on an orchard farm where his family raised sheep and chickens.

Still, even as he admitted that he might not be in the mainstream technologically, he was surprised at how little he could find on the subject of texting and driving and the law. He wasn't the only one who hadn't heard much about it. It just wasn't out there. There was one case out of New York that was vaguely related. He went to Terryl's office, and he told her he was coming up empty; the two nearly laughed about it, as if to say: Well, that shouldn't stop us.

But still, Linton needed more than he had in order to make a decision. He returned to his office. As he was searching online, he kept coming across one name: Dr. David Strayer. He was this local guy, a professor at the University of Utah, who seemed to be an authority on the risks of texting while driving and talking on the phone while driving. This fascinated Linton on two levels: A guy merely eighty miles away knew as much about this as anything else, and, more substantively, Dr. Strayer's research had showed that using a phone behind the wheel was as risky as driving drunk.

The thing that struck Linton as he reviewed Strayer's research was what was happening inside the brain. He thought: *I had no idea how much attention the mobile phone took from driving.*

"We were wondering if this was like driving while drinking a Coke, until Dr. Strayer came along," Linton reflects. "Then I realized it wasn't like drinking a Coke, it was like driving drunk, really drunk."

"The precedent that tipped it for me was not legal; it wasn't a precedent of what was going on in other states. It was a scientific precedent," he says.

Since 1990, when Dr. Strayer had begun digging into the issue of driving and attention/distraction, he'd studied the subject from numerous angles, amassing a host of papers on various nuances. For instance, in 2001 he published a paper in *Psychological Science* showing how motorists got distracted just talking on a phone; in the *Journal of Experimental Psychology* in 2003 he showed that such cell phone–using drivers don't see as much of their surroundings as drivers not on the phone. Even if they're looking at the road, he'd found, their visual acuity was impaired by the cognitive demand of the phone.

Also in 2003, he made a presentation at the International Symposium on Human Factors showing that cell phone use impaired drivers to the same level as .08 blood alcohol content, the level of legal intoxication in most states. At another conference, a year later, he showed that cell phone–using drivers are less focused on the road than drivers talking to a passenger in the car; the reason is because the passenger acts as a second set of eyes, modulating his or her conversation based on roadway conditions. Not so the person on the other end of the cell phone, who can't see what is going on. In 2007, Dr. Strayer showed that motorists using cell phones do not get better with practice.

Along the way, Dr. Strayer was doing other research, related to aging and attention, studying some of the impairment associated with Parkinson's. But the majority was on driving and distraction, and most of it was not well funded. Dr. Strayer was operating on a shoestring budget, eking out grants, and he thought he knew why: In whose interest was it to discover that there was a risk to this thing that everyone loved doing, and that was one of the most culturally celebrated activities, multitasking?

Still, Dr. Strayer had amassed a trove of data, much of it peer-reviewed and published in respectable scientific journals. And some of the most important coming from a researcher in Linton's backyard. He'd made up his mind.

Now he had to tell Daines. He hoped it would be a simple conversation, but there were layers to this case, political hurdles, just below the surface.

LINTON KNEW THAT DAINES, as an elected official, must be acutely sensitive to the community. After all, one wrong move, and Daines could face bad press, an angry electorate, and then a motivated opponent. Simple realities. That said, Linton knew that Daines had been fair and put justice first in past dealings.

An added wrinkle had to do with the Church. Reggie was LDS, like three-quarters of the just more than one hundred thousand residents of the county. That didn't mean that the prosecutors gave breaks to members of the Church. They had plenty of prosecutions to show otherwise. Terryl herself, after all, had gone after a bishop years ago.

At the same time, Reggie wasn't just a member of the Church, he was a poster child for many of its best-regarded attributes: clean-cut, churchgoing, family-focused, committed to going on a mission. He was an athlete and a decent student, things people in this county wanted their kids to be. He looked nice. He was nice.

As Linton was on his way to go talk to Daines, these things percolated. "This kid was on an LDS mission, and that is highly regarded in this valley. We knew we'd have to pull him off the mission," he recalls. "The political ramifications were very high."

Linton felt the science was so clear that it trumped any unstated risk from Reggie's Church ties. But then, Linton also had a secret reason for looking a bit differently upon the influence of the Church.

BORN ON A FARM in Utah, Linton came from a family without much education. Initially, they made ends meet in part by selling off plots of their twenty-six acres where they'd raised animals and had some orchards. Dirt poor.

Eventually, they parted with all the land and moved to the outskirts of Salt Lake City, to a poor neighborhood. Linton started to run with a rough crowd. In high school, he had a friend die of a meth overdose, and his best friend was a drug dealer—meth, marijuana, quaaludes, but mostly LSD. He once spiked Don's drink with LSD, leading to a horrible trip that left Don curled up in the corner of his bedroom, afraid he was going to fall through the floor. This was the late sixties.

Linton didn't make a habit of it. His dad told him: If you ever start taking drugs, or drinking, don't come home.

He respected his father; his dad had those big biceps. But he didn't fear he'd get beaten or anything like that. The family had a heritage of not drinking, and a religious faith. But when it came to the Church, Linton was being told one thing while having a very different kind of experience.

IT HAPPENED THE FIRST time in the church, in an isolated area. The perpetrator was a well-regarded church member. Linton was seven. He remembers the man putting a hand down his pants, and fondling him. From that point on, for three years, the man used his position and good reputation to groom Linton, isolate him, intensifying the abuse. He manipulated Linton's family into allowing him to spend private time with the boy, even overnights at the man's house.

Linton didn't tell anyone; he felt he shouldn't. He couldn't make sense of any of it, this terrible shameful thing being done to him by an authority figure, someone, no less, so well regarded in the church. Linton felt he must've done something wrong. "I felt God had deserted me and I couldn't figure out why because I'd been such a good kid."

The experience made a profound impression on him that led him to think twice about the difference between spirituality and religion, between faith and the institutions that deliver its message.

At least that's how he came to terms with all of this later on, in a rational way. At the time, he couldn't make sense of his life, of the role in it of such a prominent authority that he couldn't trust. He got depressed. He heard voices. He went on antipsychotic medications.

One thing that helped Linton find his way was music. In the fourth grade, the music instructor told Linton he ought to play the violin. He went home and his dad said: No son of mine is going to play violin. So the family rented him a saxophone, and he took right to it.

Linton became good, earning first chair in a jazz ensemble in junior high school and then high school, and winning scholarship opportunities to play music in college. Sure, he listened to Led Zeppelin and Hendrix like everyone else, but he loved Coltrane, too. Music was an escape from it all—the poverty and rough circumstances around him, the proximity of sexual abuse. Maybe he'd be an art teacher someday, he thought.

By nineteen, some of the pain subsided, at least it seemed to. Then another trauma: His beloved older sister Kathleen got very ill, first with cancer and then from the radiation treatment. He would read to her at her bedside as her condition worsened, and he prayed, sometimes blaming himself like he'd done when he was abused. "I thought maybe Kathy wasn't getting better because I'd done something wrong," he says. "No matter how much I prayed, no matter how much I fasted, she was not getting better."

"The mind-numbing, bizarre thoughts I'd experienced years earlier started to come back."

She died when he was twenty-one. At about that time, Linton went on his mission. This was what his family desperately wanted for him, and he supposed he wanted it, too. It was hard to make sense of things.

He was assigned to go to Belgium. But his first stop was Provo, the Mission Training Center, the very place that Reggie had been.

Like Baird, Linton was identifying with Reggie, but for different reasons. While Baird could identify with the clean-cut boy in Reggie, the one who had made a fatal mistake, Linton could empathize with someone who was in the middle of a mission that was about to end abruptly. In Linton's case, he'd gone to the Mission Training Center and spent three weeks there, learning French. He'd felt deeply unsettled, then depressed. And then, one evening, he got out of bed, put on his clothes, and left. No warning, no explanation. He got home and told his parents he wasn't feeling right about it. They were mortified.

Linton was taken to talk to an athletic coach at BYU, who told the young man he was a "quitter." A member of the Church told him he'd regret the decision the rest of his life.

Of Reggie, Linton says he understood: "Pulling him off his mission was going to be incredibly hard." Linton himself had left one, and knew what it was to be a young man in peril.

But Linton, while he believed in God, had lost his reverence for the Church. It was cover for his abuser. It was not sacred. And so he was free to look differently at Reggie, and the case.

It was another sign of the interesting role Reggie was playing. Depending on a person's life experience, they saw Reggie and his actions in slightly different lights. Baird saw Reggie's case one way, Linton another. Such was the razor's edge of right and wrong that texting and driving played at the time; it seemed possible to view Reggie and his actions through different lenses. Over time, Reggie would become both a lightning rod and a prism through which people in the community, prosecutors, legislators, and others around the country would view themselves and their own behavior.

WHEN LINTON TOLD DAINES of his decision, they had a decent back-and-forth. Daines expressed some skepticism, but he respected his colleague and appreciated the scientific background that Linton presented.

Linton walked out of the meeting and asked his assistant, Nancy, to put together the paperwork for the charging document.

Linton recalls thinking: "I had a lot of sympathy for Reggie. I knew this was going to be horrible for him." In the same breath, he thought that this prosecution wasn't just necessary—in a weird way, it was perfect. The perfect test case, the perfect way to set precedent—not with someone who was malicious, not with a societal outlier. "Reggie Shaw was the perfect person for me to prosecute, not because he was evil, but because he wasn't evil. I don't know how to say that. I don't know how to put that. But I knew it."

CHAPTER 29

REGGIE

REGGIE WAS FEELING HIGH a month into his mission when it came time for his first missionary conference. This was a gathering of around one hundred missionaries for a pep talk and celebration. It was held in Saskatoon, about 160 miles from Regina, a three-hour drive across the northern Great Plains. Reggie was particularly excited because of the speakers who were going to be in attendance at the retreat. They included Elder Clayton, one of the members of the Quorum of the Seventy. This was a high calling in the Church, one of the elders, figuratively and literally, who would travel to teach congregations. Also, speaking the first morning were two district presidents, and their wives. After the speeches, Reggie felt ebullient, so fired up.

He returned to his room. He was resting, getting ready for dinner. There was a knock on the door. Reggie answered.

The visitor said: "President Morgan needs to see you."

"HAVE A SEAT, ELDER Shaw."

Reggie sat in an orange cloth recliner, nothing fancy, right out of the 1980s. President Morgan stood next to a desk. He was tall with brown hair turning gray and a wide smile, though he wasn't wearing it now.

"Elder Shaw, I hate to tell you this, but you need to go home."

Reggie's eyes filled with tears. He didn't have to hear more. He had been so happy before this moment. He'd come where he wanted to be, needed to be, had put that horrible thing behind him, not just the wreck, the thing with Cammi, everything.

He stood up. He lifted the orange chair.

He threw it across the room.

He began bawling.

President Morgan let him pour out his frustration and rage. He said, gently: "The state filed charges against you. We got a call from Jon Bunderson."

The summons and charging document had been submitted on Tuesday, August 21, though word had not reached the Church and Reggie until a few days later, on Friday. The summons ordered Reggie to appear in court on the following Tuesday on the charge of negligent homicide, a class A misdemeanor. It was signed by Judge Thomas Willmore, who had drawn the case. He'd been appointed to the bench in 1999 by then governor Michael Leavitt, a Republican.

President Morgan tried to calm Reggie, but the young man was inconsolable, turning inward. President Morgan gave him his instructions: Return to Regina, pack your things. Tomorrow, you go home.

AT THE APARTMENT, REGGIE called his parents from the landline. Nobody answered the phone at the Shaws' house. He tried his mom's cell phone, but she didn't answer. He called his dad's cell.

"Who is this?" his dad answered.

"It's your son, Reggie."

"Oh, sorry. We're at the high school football game."

Reggie felt a moment of anger. *How could you be at a football game?* He even asked.

"It all happened so fast, Reggie. They decided to press charges, negligent homicide."

It was just like his dad, Reggie thought. To the point, matter-of-fact, life will go on and we'll be fine. More likely, Ed was just overwhelmed, paralyzed. It was his greatest terror, the idea of his son in jail.

"They pulled the phone records," his dad told him. "They said that when the accident happened, you were texting."

"There's no way that happened."

"Reggie, we'll pick you up at the airport tomorrow."

"There's no way that happened. That's not what happened," Reggie said.

He could hear the football game in the background. This wasn't happening.

"We'll get you at the airport."

They hung up, and it crossed Reggie's mind that while his parents would be at the airport, he might not be. He could take off his LDS badge, change into street clothes, and get a job in Canada.

A FEW DAYS LATER, back home from a mission for a second time, Reggie couldn't eat. It was virtually unheard of for a missionary to return home twice. His stomach was in knots. He got ulcer medication. His mom took him to see Gaylyn, the longtime family friend and counselor.

Gaylyn saw in Reggie a person who seemed at his wit's end. She asked him to fill out two questionnaires, the Burns Anxiety Inventory and the Burns Depression Checklist.

The tests had questions you answered on a scale of zero to three, with zero meaning something wasn't a particular concern and three meaning the client was bothered "a lot." Anytime a client would notch a two or three (moderate or a lot), Gaylyn paid close attention.

Among Reggie's scores:

Fear of losing control—moderate, a 2.

Fear of cracking up or going crazy—moderate, a 2.

Fears of criticism or disapproval—a lot, a 3.

Pain, pressure, tightness in the chest—3.

In all, on the anxiety test, Reggie's answers added up to a 35. That was a red flag, for sure. Any score over 31 signified "severe" anxiety.

On the depression test, more of the same:

Was he feeling sadness, down in the dumps?—3

Low self-esteem—2

Are you self-critical, do you blame yourself?—3

Have you lost your appetite?—3

"Reggie, you are in a severe stage," she told him after reviewing the numbers. She recommended that he get a prescription for an antidepressant.

He drove home, drained but not feeling any better. Later, he and his mom got a prescription from a local doctor. But when he got home with the prescription, he went out back of the red-brick house. He pulled the prescription out of his pocket. He tore it up. He put it in the dumpster.

He thought: *This is something I can handle. I can deal with this.*

And there was something else going through his mind: "The pills would make me feel a false sense that everything was okay when nothing was okay. The pills would make me feel like everything was all right, and I could just live every day like they were all right."

He knew his mother would find out from Gaylyn he'd gotten medication. So he told her he'd filled the prescription and was taking the drugs. Another little tiny lie, right? Would it be harmless?

REGGIE, AND HIS FAMILY, met with Bunderson. What the lawyer heard, overwhelmingly, was that the family wanted to fight. Reggie, too, but Bunderson's impression was that the family wanted it more. Regardless, he started putting together ideas for a defense. His impression, on its face, was this was a "tryable case," one he had a reasonable chance of winning.

As Bunderson sought information, he was curious about Reggie himself. Exactly who was this kid? How might a jury react to him?

"Let's go back," Mr. Bunderson said. He wanted to talk about what happened before the accident.

Before?

"Your mission. What happened there?"

Reggie fidgeted.

"I had an issue with a girl," Reggie said.

Recalling back, Reggie reflects: "He gave me a look."

"I need to know the details," the lawyer said. He asked Reggie whether he'd feel more comfortable if his parents left the room. Reggie said they could stay. He explained about how he'd lied to Bishop Lasley, telling the lay leader that he'd not had sex with his girlfriend.

"He was trying to figure out how I lied and why I lied and if something would ever come up in court about how I lied before about big issues."

The stuff with the mission, questions about his own character, in his own mind, weighed heavily on Reggie, more than he let on.

"That was a character problem, and I realized that, at that point, it could affect potentially more than my mission. It could affect anything I had to say in court."

Bunderson realized it, too. It could be a huge problem. Assuming, that is, the prosecution convinced the judge to allow such evidence.

"If the prosecution were to tell a jury here in Utah that he'd lied to his bishop about something he'd done before going on a mission, then they might think he's lying about everything."

But as Bunderson thought through the possibilities, he figured there were some very good ways around that ever happening. Yes, this case was winnable on that front, and a lot of others.

CHAPTER 30

HUNT FOR JUSTICE

O N THE MORNING OF September 18, 2007, Leila O'Dell put on a black floral skirt and a matching sweater. She drove her Prius to the Cache County courthouse.

In contrast to the more traditional surrounding architecture, such as the county attorney's office across the street, the three-story courthouse sports steel trim around a checkerboard of clear glass windows that stretch the building's height. Inside, light bathes the hallways. The sun, or flat light on a cloudy day, pours in from every window; if there are secrets here, it seems to say, the light will quickly find them.

Leila took the elevator to the third floor, and entered courtroom 5. It was large and packed; this was not just a hearing for Reggie but for many others being arraigned. She found Megan, and Jackie, and Terryl. They sat in the long brown courtroom pews, taking in the light through the south-facing windows and views of the three sets of jagged-topped mountains reigning over the three canyons beneath them—the Logan, Providence, and Blacksmith Fork peaks.

Linton sat at the front. He got his first look at Reggie in person. He was struck that there was something different about this kid from virtually every other defendant in the place. No tattoos or piercings, no shifty or menacing look. Something all-American about this guy.

Megan O'Dell glared at Reggie and thought: *You killed my dad and I hate you!*

For Leila, sniffling, holding it together, it really wasn't about Reggie. It was about Keith. Regardless, she also believed that this whole thing, the entire deadly affair, was about to reach some kind of closure: Reggie was going to plead guilty, of course, she thought. The evidence with the text records was, as they say, black-and-white.

Judge Willmore also saw Reggie for the first time. The judge had grown up in Logan and graduated from high school there; attended law school in Sacramento, California; then returned home and practiced law for sixteen years before being appointed to the bench where he'd presided for nearly seven years. In his chambers, his bookshelf kept histories of great men, like Ben Franklin, Harry Truman, and, most cherished by the judge, Lincoln. Two books were more cherished than the others, and he kept them in his desk. One was a big blue manual for Alcoholics Anonymous, something he consulted often because so many cases he presided over involved addict defendants. The other was a beat-up paperback edition of Victor Hugo's *Les Misérables*, in which he'd underlined passages in red and blue pen. It touched on so much of what he faced: crime, rehabilitation, the role of the system for good and sometimes ill.

He'd gotten the *State v. Reggie Shaw* by lottery, a random assignment of cases to one of the four judges in the district. A veteran, he approached this stage of the process, the arraignment, with efficiency. Meaning: He didn't get too invested in individual cases, or even wish to be particularly informed about them, because there was no point in doing so. He didn't know if someone would plead guilty, not guilty, or reach a plea. He and others in the system referred to this mass arraignment as a "cattle call."

"How does the defendant plead?" he asked when Reggie's case came up. The judge had short, manicured hair, slightly graying, and he spoke quietly in a way that suggested he was restraining an intensity.

Bunderson responded: "We enter a plea of not guilty and request a preliminary hearing."

Leila was incredulous. "You're kidding me. You can't be serious!" Her mind was racing. This was supposed to be so straightforward.

The judge set a date.

Leila and the others were struck by what she called Bunderson's "belligerence." But he was doing his job. They might well have been reacting to what they saw as a travesty of justice. It was the first outward sign he was going to fight zealously for his client in what he saw as a winnable case.

JUST AFTER THE HEARING, the families—Leila and Megan and Jackie—along with Terryl and Linton, went down the hall in the courthouse to a conference room. They had a short meeting, where Linton tried to set expectations. He told them that jury trials can be unpredictable, even though he thought they had all the facts and the law on their side.

He gave them a 30 percent chance of prevailing. He was trying to set realistic, if not conservative, expectations.

He was not surprised that Bunderson was pushing the issue. He suspected that a different defense attorney, maybe someone the Logan prosecutors knew better, might have taken a quick plea bargain. Not Bunderson.

Linton had already begun preparing for a full-blown trial. On August 15, about a month earlier, he'd written a memo to all the attorneys in the prosecutor's office. It was titled "Texting While Driving."

"Last night I found out that on August 7, 2007," he wrote, "a Harris Poll was released showing that 89 percent of American adults believe that sending text messages while driving should be outlawed."

He went on to say that, despite this awareness, "amazingly," 64 percent of the adults in the survey said they had sent a text while driving.

Linton was thinking: People did it, but they knew it was wrong. If they looked inside themselves, they could see the disconnect. They'd sense that Reggie should have known the risks of texting and driving—that he must have known.

He concluded in his memo: "This might be an important issue in any trial conducted in *State v. Reggie Shaw* and how we choose the jury and approach the issue."

THESE SAME ISSUES WERE going through Bunderson's mind, too.

His first thought was: Who knew?

"Who knew this was grossly negligent conduct?"

He was thinking: Who had told the public you weren't supposed to text and drive? Where were the studies? Who told Reggie?

A second big issue for Bunderson had to do with the evidence that Reggie was texting at the time of the accident. Yes, the phone records clearly showed some texting activity. But they didn't prove that Reggie had been texting at the exact moment of the accident.

He thought: They won't prove that he was texting at the exact moment.

And he felt a jury would be much less inclined to convict Reggie if the best the prosecution could do was show that Reggie was texting thirty seconds before the wreck. If Reggie had been texting thirty seconds earlier, Bunderson thought, how could that have anything to do with the accident?

Bunderson was putting together other defenses. He was looking at investigative misconduct, small and large. Could he find problems in the investigators' pasts, something that might call into question their credibility or capability?

On September 11, he wrote himself a memo indicating he'd spoken with Briana Bishop. She'd told him about her two interviews, the first one at the law enforcement offices, but the second time at her work. He writes: "The second time one of them came to her work, (and) embarrassed . . . her."

In addition to these smaller things, there was one huge issue on the investigative front. It was a question of whether Rindlisbacher had mishandled his initial contact with Reggie. Bunderson wondered whether Rindlisbacher's testimony could be tossed out because he hadn't read Reggie his Miranda rights, the warning that informs a criminal suspect in custody that he or she has a right to remain silent and has a right to an attorney.

There was no disputing that Rindlisbacher had not said this to Reggie. But, to Bunderson, he should have. When the trooper took Reggie to the hospital, "I thought that Reggie was in custody and interrogated," Bunderson says, even though at the time Reggie had not been arrested or charged with a crime.

If he could show Reggie should've been given such a warning, then he could have Rindlisbacher's testimony tossed. But that wasn't really Bunderson's aim. He had a larger goal. He wanted to keep Reggie from having to testify.

Since the only statements Reggie had given were to Rindlisbacher on that day, they would be the only statements that Reggie would have to personally address to the jury. Without Rindlisbacher, there would be no Reggie. And that meant no tough examination of what he remembered from the accident, no asking him to explain the phone records, no asking him what he knew about the risks of texting and driving, and no asking him to reveal why he'd left his first mission. Reggie, nice and quiet though Bunderson thought he was, wasn't someone the lawyer wanted to see testify in his own defense.

The sum of all these strategies that Bunderson was thinking through added up to one more tactic: stalling. The longer they could draw this case out, the more it might find its way to the bottom of the pile of prosecutorial priorities, and the more the public concern about two dead rocket scientists could ebb. Bunderson made no bones about it in talking to Reggie and his family.

"It's rarely in a defendant's advantage to hurry things up," he said. "The longer you give for the steam to go out of a situation, especially this one—the deaths got a lot of press, folks were reading the newspapers— the more they would tend to forget all about it."

He told the Shaws: "Let's not be in a big hurry."

Each passing day seemed to fuel the other side, anxious for closure. This was particularly true of Terryl, who kept digging and digging, seeing a great injustice that stretched beyond Valley View Highway on the morning of September 22, 2006.

ON NOVEMBER 19, AMID the back-and-forth motions and delayed trial dates, there was a testy interview that symbolized not just the antagonistic tenor of a case that would drag on for well more than a year but also highlighted the ways in which society at large was struggling with how to think about our relationship to technology. What is okay? What can be proved? What steps are we willing to take to protect ourselves and our own from truth and consequences?

Singleton and Rindlisbacher were interviewing Mary Jane Shaw. It was 9:31 in the morning. Mary Jane was joined by Bunderson and her son, Phill Shaw. Both sides came in with guards up; the family thinking Rindlisbacher overzealous. Separately, there was civil liability, with Phill worrying "they could take everything," if the Furfaros and O'Dells successfully sued the Shaws.

The investigators, meantime, saw the Shaws as meddling liars.

Not only were the sides confronting each other, the person at the center of the whole thing was Mary Jane, Reggie's mom. His fierce protector.

That was the context going into the interview, which Singleton would later describe as "the most contentious I've ever had in twenty-four years in law enforcement."

IT STARTED POLITELY ENOUGH. It took place in a small conference room on the top floor of the county attorney's office. Singleton recalled that, before the official interview took place, Mary Jane had mentioned to Rindlisbacher, "I saw you at the temple the other day." Singleton took it as light code for "We're pals, and I'm a good upright LDS person." Singleton was LDS, too. Mrs. Shaw answered Rindlisbacher's questions about the weather, saying it was "horrible." She talked about finding Reggie at the scene, how she hugged him and asked if he was okay. Toward the end of Rindlisbacher's questioning, things began to get testy as the trooper asked whether Mary Jane had told Reggie not to cooperate with the police. She said no, she hadn't done that.

Then things really intensified when Singleton took over the questions.

"Walk me through the events of September 22, 2006. Start from—"

Phill cut her off. "Objection. Asked and answered. She only has to answer these questions one time, Singleton, and you've already tried to ask her by going over them. Ask—"

Singleton: "You can either butt out—"

Mary Jane: "It's okay."

Phill: "Don't give your opinion here."

Singleton: "I'm not here to talk to you."

Phill: "Don't ask questions that she's already answered or she will walk."

AS THE INTERVIEW CONTINUED, there emerged all the contours of the battle, the idea that Reggie's the victim of zealous prosecutors, a veritable witch hunt. The suggestion he'd been a liar, with his family as co-conspirators, egging him on with his deception and denial. But it was a bit later that, with equal intensity, the conversation turned more directly to texting and driving, and whether Reggie was doing it and whether, regardless, it could be proved he was doing it at the time of the wreck, and had caused the wreck.

These exchanges are important because they frame a debate that was never so succinctly argued in the courtroom, given the long and drawn-out legal process. This was, in a way, a proxy for some of the reasoning in a jury trial that kept getting delayed and delayed, and a proxy, too, for the larger policy issues that allowed society to keep at bay the simmering legal and policy conflicts and challenges.

Singleton: "The cell phone records indicate, through the investigation, that Reggie was text messaging at the time of the crash."

Phill: "Objection. Do not answer this question. Speculation. They don't know the exact time of the crash. They do not have phone records to show if he was text messaging at the exact time of the crash because they do not know the exact precise second of the crash. Do not answer this question."

Singleton: "I'm asking if cell phone records show the crash happened at that time—"

Phill: "Objection. No foundation. Don't answer this question."

A little later in the interview, the subject came up again, this time with the question being, again, what Reggie had told his mother about texting and driving. The questions now were coming from Rindlisbacher who, curiously enough, had taken on a bit of the good-cop routine, with Singleton asking the more pointed questions. In answer to Rindlisbacher, Mary Jane said Reggie had always maintained he was not texting.

"He's never, ever, in a year and how many months, said anything different."

And a bit later she said: "He said no, he was not, at the time of the crash."

"At any time?" Rindlisbacher asked.

"I didn't ask him at any time. I asked him, when the crash happened, were you on the phone? He said no."

For Phill, there was no lingering doubt, not exactly. But he was vaguely aware of the complicated psychology at work. Reggie maintained he hadn't been texting. Phill also didn't personally look at the texting records. And, looking back, he felt that, for all of Reggie's assurances that he hadn't been texting, Reggie might have been taking cues, too, from the family. The intensity with which the family undertook the defense had a self-perpetuating and escalating force: Reggie denied texting, the family backed him up, and Reggie, never someone to let others down, dug in deeper.

"He was kind of quiet, and then everyone took over. Me, my mom and dad, all in his defense, and he just went with it," he said.

CHAPTER 31

HUNT FOR JUSTICE

January 18, 2008

Dear Ms. O'Dell,

    I felt sad to read your letter and recalled the tragic incident that took the life of your husband and coworker. I have only followed the case in the news and I think the filing to the charges under these circumstances was somewhat new.

Thus began a letter to Leila from Utah state senator Lyle Hillyard, a Logan lawyer and one of the most powerful state senators in Utah. He was writing in response to a letter Leila had written to him a few weeks earlier, asking the senator to take on texting and driving.

In his reply, Senator Hillyard wrote that the January session was coming up in only three days and therefore he wouldn't be able to do anything about the issue this session. But he said he would refer it to his staff for study to see if it was worth undertaking a year from then.

You raised a legitimate issue that I think the legislature needs to look at seriously. We all need to remember that when we drive a car very serious

consequences can occur if we are not paying our full attention to what
we are doing.

JUST A FEW DAYS later, state representative Stephen Clark was driving up I-15, the big highway that runs through Utah. He was heading from Provo to Salt Lake City, the very drive, in fact, that Reggie had made with his parents after his first mission came to an abrupt end.

Traffic was bad and Clark was rushing. He was chair of the appropriations committee in the Utah House of Representatives, where he'd been first elected in 2000. A contractor by trade, he focused on commercial infrastructure projects, like plumbing and heating; he was a Republican, like most of the legislators, but a moderate one. Meaning: He was dyed-in-the-wool on issues like abortion, and was a fiscal conservative, of course, but he was willing to consider less strict rules on things like immigration. He'd tried, without success, to get a law passed to allow immigrant workers to get permits to work in the state.

As he drove up I-15 in his 2002 gold Lexus, he shook his head with frustration. He was going to be late for the legislative session. He pulled out his phone and texted his secretary.

"I was late and traveling fast and the traffic started backing up," he says. He finished his text. When he looked up, he discovered the car in front of him had come to a stop. And he was about to slam into it. "I slammed on my brakes and just barely missed ramming him."

He was mortified by his own behavior. He felt an almost immediate sense of resolve. Despite his history of fighting government regulations, he decided he wanted to do something about texting and driving. "This has got to stop," he decided. "It's got to stop not only with me but with those who are going to be behind me, texting."

Something else, he said, flashed through his head or, rather, someone else. He thought about this kid he'd heard about, Reggie Shaw. He'd read about Reggie when the accident happened, and he thought about that accident now and again as he'd watched his own behavior and that of others on the road. As he continued his drive to the state-house, he says: "I thought to myself: I could be another Reggie. I could be in a situation like that, or I could be like those rocket scientists."

But there were some realities to contend with. One was that it was too late to introduce legislation for the 2008 session. More worrisome over the long run was the reality of Utah politics. This was a place where a conservative like Clark could be seen as moderate, in a deeply red place, where "government interference" were fighting words. "The Utah legislature is very conservative," Clark says. "They don't like the government telling them what to do."

CHAPTER 32

HUNT FOR JUSTICE

O N THE LEGAL FRONT, the late winter and spring brought a flurry of motions and countermotions that spoke to the specifics of Reggie's case, but also to broader issues as the legal system wrestled with technology, law, and the brain.

On January 24, 2008, Bunderson filed a motion to prohibit the prosecution from introducing diagrams of the crash site, excluding any diagram "prepared by Mr. Kaiserman, the only eyewitness."

The motion asserted that "there was no debris found, there were no skid marks, scuff marks, or gouge marks, and no evidence which could support the preparation of a diagram."

The essential point Bunderson was arguing was that there was no way to show precisely where the accident had taken place, on which side of the yellow divider, and, if Reggie had swerved into the oncoming lane, it could not be proven how far he'd swerved.

JUST TWO WEEKS LATER, on February 8, Bunderson asked the judge to rule that he would advise the jury to question the reliability of the farrier. "Mr. Kaiserman had only a fleeting moment, if not just a split second, to observe the situation, and he was involved in a major collision himself," Bunderson wrote in his motion.

That day, Bunderson filed a handful of motions, the sum of which constituted the essential pieces of his case. One motion went to the heart of what everyone saw as a key issue, Reggie's character. It was just one page, promising supporting documentation to follow, that referred to Reggie's history with the LDS mission. It said that the discovery in the trial "vaguely references" the fact Reggie had gone to the Mission Training Center and then returned. That fact, Bunderson argued, had no bearing on the accident and, borrowing from the legal vernacular, was too "probative," meaning it could carry more weight in the eyes of the jury than was relevant to the case.

Another February 8 motion was aimed at disallowing the prosecution from using anything in the trial that Reggie had said to Rindlisbacher at the scene of the accident, and en route to the hospital. This had to do with Bunderson's assertion that Rindlisbacher failed to and was required to read Reggie his Miranda rights.

This motion wasn't just aimed at limiting Rindlisbacher's testimony. It would serve the perhaps more important aim of keeping Reggie off the stand. After all, if the prosecution could introduce nothing from Reggie's own mouth, then the defense would not have to devise a strategy for refuting it. It would be easier for Reggie to assert his right not to testify in his own trial.

Also on February 8, Bunderson filed a motion that highlighted some of the new issues raised by the *State of Utah v. Reggie Shaw*. The motion had to do with what was going on inside Reggie's mind, and it raised fascinating questions about the collision of technology, science, and law.

The motion sought to disallow the testimony of David Strayer, the professor at the University of Utah, who is expert in distracted driving. Bunderson wanted to avoid Dr. Strayer testifying to the jury that when a motorist is texting, or, for that matter, talking on the phone, they are unfocused and distracted. Bunderson argued that Strayer would be speculating about Reggie's "state of mind," something he argued is not allowed. In turn, Bunderson said, Dr. Strayer, an "expert," would be telling the jury that Reggie was negligent, taking that decision away from the jury.

He wrote: "The opinion of an expert that texting or using a cell phone is distracting or even dangerous is nothing other than telling the jury that such activity is either negligent or criminally negligent."

IN THE PROSECUTION'S RESPONSE to the motion, Linton conceded one point: "With respect to a defendant's state of mind when he committed an act, an expert cannot testify that a defendant acted with negligence (or for that matter, recklessness, knowledge or intent) because such an opinion would constitute a legal conclusion . . ."

But he also wrote that Dr. Strayer should be allowed to discuss the risks of texting.

"He is expected to testify that text messaging can cause a person to focus on things other than driving, and that at high speeds driving while texting is consistent with driving over center lines, and in general, erratic driving."

Linton argued that the Utah Supreme Court had ruled that experts were allowed to testify about how certain acts are "consistent with" certain behaviors. For instance, he cited one sexual abuse case in which the Supreme Court had found that a doctor could testify that a sexually abused child could exhibit "sleeplessness, poor appetite, fear of certain individuals, clinging behavior, and urination problems."

Linton argued that the Supreme Court found that such testimony didn't necessarily say that a particular alleged victim was abused but, rather, that such behaviors could fairly be raised as evidence of such abuse.

"While Dr. Strayer's opinions will touch upon the ultimate issues of this case, they will not include a conclusion by the doctor that the defendant was negligent."

Linton had filed his response on March 17. A few weeks later, in early April, Bunderson fired back with a three-page motion that put a fine point on his argument about distraction and frame of mind.

"Focusing on things other than driving, and erratic driving, are the very definition of negligence," Bunderson wrote.

And, he wrote, "this case is unique because the mental element is criminal negligence, and for an expert to testify that text messaging causes someone to fail to pay attention is, at the very least, testifying to an inference of a negligent mental state."

What Bunderson was getting at was one of the essential aspects of the case against Reggie but also about the nature of texting or using a phone while driving: Is it inherently distracting? If that is the case, he thought, and the jury believed that the very act was distracting, then it might determine that Reggie was inherently negligent for having done it.

That was very different, Bunderson thought, from, say, eating, which no one was arguing was inherently distracting, not creating a cognitive load. Sure, of course, someone could get distracted by reaching for food, looking away from the road, or even by turning the dial of a car radio. But those were momentary acts, not systemically provable acts of distraction.

Bunderson didn't yet know the neuroscience, but he understood the risk: If an activity was inherently distracting, then it could be seen as inherently negligent.

He didn't want the jury to hear that at all.

ON OTHER POINTS RAISED by Bunderson, Linton argued some and conceded others. For instance, the prosecutor agreed that there was no reason to bring in Reggie's failed mission, and he had never intended to. He wrote in his own motion on this issue, "The state of Utah does not know why the defense believes this.

"Had the defense called the state, the state would have stipulated not to introduce any information regarding the fact that the defendant had to come home from his mission for personal reasons once, and then a second time for the charges in the case. The state agrees that this is not relevant."

Notably, Linton had referred to Reggie returning from a second mission. Bunderson might not have wanted to have Reggie's first failed mission admitted, but at the same time, the state didn't need the jury to hear that Reggie was so powerfully committed to going on a mission that he tried twice. If not for the state's zealous pursuit of Reggie, he would be serving the Church. The character issue could work both ways, and now it was coming out of play.

ON THE WAY TO each court hearing, the Shaws avoided Valley View Drive. They never drove to the crash site, or directly past it. Reggie still hadn't been back to the site. It was the elephant on the road, and it would remain unacknowledged.

On every trip to Logan, the family did, however, visit the city's majestic temple, the second oldest in Utah, right up against the mountains, viewable from the Cache County Attorney's Office. While she prayed, Mary Jane took to thinking about the Christmas song "I Heard the Bells on Christmas Day," based on an 1863 poem by Henry Wadsworth Longfellow.

The verse that struck her had to do with hate:

And in despair I bowed my head;
"There is no peace on earth," I said;
"For hate is strong
And mocks the song
Of peace on earth, good-will to men!"

Mary Jane prayed not to hate—not to hate the system, not to hate people who were saying things about Reggie and what a horrible person he was. She begged God for help.

"Bless us to get through this," she prayed. "Bless us that those who are trying him will see what's in his heart."

CHAPTER 33

TERRYL

IN JANUARY, JACKIE GOT a visitor from Indiana, Gary Maloney, the longtime family friend who'd been the best man at her and Jim's wedding. And, lately, a virtual friend.

Jackie and Gary had been bonding for months in World of Warcraft. They played after the girls went to bed. They'd pair up their characters: When she played with Moonrise, her night elf hunter, he played with Pam, a dwarf hunter; when she was the human priest she'd named Serifim, he played a night elf he named Tsparhoc. They went on quests and explored dungeons, and they chatted.

When he came to Tremonton in January, visiting family in the area, he spent an evening hanging out at the Furfaros'. He and the family sat in the living room, watching *Battlestar Galactica*, and something special happened. Cassidy, the youngest girl, went over and sat with Gary. Stephanie then did, too.

"It was like a first moment for me," Jackie says. "I could be with this guy."

In April, after many more nights of online interaction, Gary decided on another visit in May. Jackie told the girls, and they asked: "Where is he going to sleep?"

"Probably on the couch," Jackie told them.

"Why not the bedroom?"

They obviously were just trying to make sense of what kind of friends slept with what kind of friends. She explained about how married people share beds. "We're not that kind of friends," Jackie said.

Gary came in May, around his birthday. Up until that point, Jackie hadn't had much of a social life, not a single date. She got a babysitter, and for Gary's birthday, they went to an Olive Garden and then to the movies to see *Indiana Jones and the Kingdom of the Crystal Skull*. Then they went home and sat on the couch and put on *Lara Croft Tomb Raider: The Cradle of Life*. Media, movies, video games, television were their common interests.

They sat there quietly, each trying to figure out how to broach their mutual affection.

Finally: "I was wondering if we could be a couple," Gary asked her.

"Yes." Jackie's reaction was immediate and succinct.

That summer, things blossomed further when Gary came for a weeklong visit while the kids were with their grandmother. They got some alone time, and watched a bunch of movies. For Jackie, their friendship seemed like a version of a real-world happy ending, a realization of her aspiration to move on from Jim's tragic death and provide a stable life for the girls.

She told herself she didn't feel guilty. But that June, she began having vivid dreams. In one, she was hanging out with Gary, and Jim showed up on the doorstep.

"I'm not dead," he told her in the dream.

Still dreaming, Jackie told him: "That cannot really be."

FOR THE VICTIMS—AND THEIR advocate in Terryl—the year 2008 evolved into one of living with uncertainty, under the haze of an amorphous legal cloud, and an emotional one. They were adjusting to life without Jim and Keith. If Jackie was seeing some light at the end of the tunnel, that was less true for Leila.

Her life was a blur of grief and humdrum, punctuated by periodic court hearings. She went back to work as a bookkeeper, keeping the same schedule she had when Keith died, working Mondays, Tuesdays, and Thursdays. Her income was modest; she made between $25,000 and $30,000 a year, and she now had to grapple with a new bill, health care insurance for her and Megan, something Keith's job once covered. The policy cost her $400 per month initially, then leapt to $850.

Aggravating the costs was Megan's ongoing struggle with severe kidney stones. More than once, the young lady called her mother in the middle of the night in pain, and the pair rushed to the emergency room.

She felt her relationship with her daughter was decent, even improving. Megan would come over regularly for dinner. One of Leila's pleasures was cooking, lots of hearty soups in the winter, and salads as things got warmer. When Megan wasn't there, which was most nights, she'd cook for herself and then sit down with a book and read with the television going in the background. She'd watch the news and stock shows that bore increasingly bad news about the economy as 2008 wore on. She tried to keep track of her and Keith's once comfortable savings, which she had hoped to not touch until later but now was forced some-times to dip into. She watched police dramas, *NCIS* and *Law & Order*, and she popped back and forth with her book, having to go back and reread parts she'd missed when focused on the TV.

It was television, she thought, that brought out the rare hint of judg-ment from Keith. Not the medium, but some of the people who were on the shows, like Jerry Springer. They were all the things he tried not to be, showy and self-referential. "'Who are these people seeking atten-tion?'" she recalled him saying. "'Who on earth would go out there and act that way and air their personal issues to the max?'

"He was an introvert to the max. I'm an introvert, and he was more of an introvert than me."

It explained in no small measure the depth of her grief. She'd found someone she could share the world with, quietly. She wasn't comfortable putting herself back into the world. And if she didn't do that, could she ever share herself with someone else?

OF THE TWO WIDOWS—JACKIE and Leila—Terryl really connected to Jackie. The pair were becoming friends. It wasn't so much that Terryl didn't connect to Leila, but, rather, there was just less chance to do so. Their lives didn't naturally intersect quite the same way, and Leila's introversion didn't invite much else.

In Jackie, Terryl particularly appreciated someone who looked at the glass as half full, and who forged ahead with her life—in Terryl's own mode.

But as their relationship strengthened, there was one place Terryl harbored a quiet reservation about Jackie. Terryl didn't understand her family's heavy media use. Well, more than that, she was silently critical of it. She worried that Jackie's girls would get lost in media or escape to it. Her observation came at a time when media use in general was exploding among children, up to an average of around ten hours a day of screen time, according to the Kaiser Family Foundation.

Broadly, Terryl associated heavy media use with a lack of engagement between parents and children, and less emphasis on school.

Jackie felt her girls were doing fine and that media use was part of the family dynamic. And the girls, Stephanie and Cassidy, were doing well in school, particularly in math. Besides, Jackie pointed out, Utah was cold a lot of the year, and computers, television, and video games were a better way to spend time than "sending them outside when it's thirty degrees or less, which is not practical."

Jackie conceded that some of the television shows the family watched together, like crime dramas, "are probably not great," and could be "kind of gruesome." But homework always got done, and the girls spent plenty of time with their friends.

Later research would show that Jackie's views are consistent with the perspective of the majority of American parents, according to a survey published by the Center on Media and Human Development at Northwestern University. The 2013 survey found that 59 percent of 2,300 parents surveyed did not worry that their kids would get addicted to technology, and 78 percent said they do not have concerns or conflicts inside the family around media use.

The responses surprised one of the study's authors, Vicky Rideout, a pioneer in the study of children's media use. But there was another finding in the survey that perhaps explained why parents were not all that concerned: They themselves were massive consumers of media. In nearly 40 percent of families, the parents consume eleven hours of screen media a day, which does not include work time. (The survey double-counts the time that a person uses two sources of media at once; for instance, if a person spent one hour simultaneously watching TV while surfing the Internet on a laptop, that counted as two hours of media time.) And in another 45 percent of families, parents consumed five hours of screen time per day. These figures "startled" Rideout, but she said they also helped explain why parents weren't worried about their children, given that the parents were setting a tone and the children were following.

This perspective in general—the lack of concern among many parents about the explosion of screens and screen time—drove, to some extent, an interesting change in policy in 2011 from the American Academy of Pediatrics. Recall that, in 1999, when Terryl had cut her cable television, the pediatric organization had called on families to prevent children under two from watching any television.

But in 2011, the group liberalized its policy, urging parents to limit and monitor toddler screen time but not calling for such a rigid ban. Between 1999 and 2011, the research had begun to mount and solidify about the risks of heavy screen time, but the pediatric association felt it needed to be realistic, as Dr. Ari Brown, an author of the revised policy told the *New York Times*.

"We felt it was time to revisit this issue because screens are everywhere now, and the message is much more relevant than it was a decade ago," Dr. Brown told the paper, which went on to write: "Dr. Brown said the new policy was less restrictive because 'the Academy took a lot of flak for the first one, from industry, and even from pediatricians, asking, "What planet do you live on?"'"

ON THE PERSONAL FRONT, Terryl and her family were enjoying more good fortune during the first half of 2008. In the spring, Taylor and Jayme had jointly entered the National History Fair with an exhibit on a writer named Malika Oufkir—the daughter of a Moroccan general—who was imprisoned for years in brutal conditions. She went on to become a rallying point for prisoner's rights, and the subject of an exhibit by the Warner children that won the school, regional, and state competitions, meaning that, for the second year in a row, the clan went to the nationals.

Of course, there continued to be occasional downsides to Terryl's optimism, pep, and conviction that she'd do anything to connect with her children in a way she hadn't felt connected to as a child. Terryl could overcompensate with the passion of a convert. It happened again in May of 2008, when Taylor and Jayme got a zip line, which was a wire you attached across an open space, like a yard, suspended from two high points. Then you'd slide down the wire holding on to an attached handle.

Jayme and Taylor built the line in the backyard. But it looked like it might be dangerous. Jayme took a try and narrowly avoided a crash landing. Taylor was too scared to try it. "Oh, c'mon, you guys, I'll do it!" said Terryl.

She jumped off the deck, and, halfway across the yard, slammed into a planter box. Terryl broke her ankle. "I was laughing and crying at the same time."

"My mom has an extreme sense of enthusiasm," Jayme says. "She's willing to do or try anything."

CHAPTER 34

REGGIE

DURING THE FIRST HALF of 2008, as the legal drama hummed in the background, Reggie remained folded in on himself. He lived with his folks, staying in his childhood room. He detailed cars and was the assistant coach on the sophomore basketball team. He dated a bit. And he ran, lots of running, pretty much daily, often a loop of seven or eight miles that took him from the house to the McDonald's. On an early MP3 player, he listened to rap and hip-hop.

It let him push down the guilt and fear. "It was the only thing that kept me sane." It let him keep at bay, in particular, the terror that he'd wind up in jail, and the television-inspired nighmares of what it would be like, the idea he'd have to fight. He'd had a few scuffles before in high school, glorified shoving matches between his group, the jocks, and the cowboys. He'd have regular visions of the door to the jail shutting behind him.

He did something he hadn't done much before in his life: got angry with his mom. At some point in virtually every day, she'd ask how he was feeling or say, "I talked to Jon today," referring to Bunderson, or "The next time we go to court is . . ." Reggie would tell her he didn't want to talk about it, or be dismissive, and his disrespect for her would add half a teaspoon more of shame.

In March, Reggie decided to change his circumstances. He made up his mind that he would move to Salt Lake City and attend Salt Lake Community College. He found a job working twenty to thirty hours per week, earning $8.50 an hour, for the Salt Lake Bees, a minor league baseball team. To trek back and forth to Tremonton, and around Salt Lake, he spent $3,200 on a 2003 Chevy Cavalier. He bought the car entirely on credit, with payments of $75 a month.

ON APRIL 15, THERE was a court hearing that started with a request from Bunderson that the state of Utah pay for Reggie's defense to hire expert witnesses.

Bunderson argued that Reggie was indigent, broke. This led to a relatively spirited back-and-forth between Bunderson and Linton, with Reggie taking the stand at one point. He answered questions from Linton and Judge Willmore in clipped language, polite on its face, but distant and just shy of hostile. On the stand in jacket and tie, his head seemed to hang heavy, even his eyelids hung, as if whatever was inside of his brain—unspoken thoughts and emotions—were casting a shadow over the witness box.

Linton and Judge Willmore pressed him on whether he could be indigent if he just bought the Chevy Cavalier. They asked what he'd done with the $7,500 or so his records showed he'd earned since getting back from his mission. Gas, he said, going out to eat, and dates. He said he had a savings of $90.

Judge Willmore pressed him on who would pay for college and when tuition was due.

"I ain't sure when the deadline is for tuition," Reggie responded.

In the end, Judge Willmore decided to have the defense and state split the cost of the witnesses.

There were larger issues betrayed by the testimony that day. Reggie's lack of emotion, appearing as arrogance or even hostility, fueled the prosecution. Linton's polite but point-by-point challenges, while typical of a litigious process, fueled Reggie's team's sense of indignation and martyrdom. The legal game was on now, taking on its own life.

Relations between Linton and Bunderson had frayed. Reggie and his team felt there was no basis for these charges—texting hadn't caused this wreck and, regardless, it couldn't be proved. This was an accident and a witch hunt; the prosecution felt the defense had insulated itself from science and reality. It was a real, honest, deep-seated divide.

Each side wanted to win and felt it deserved to. But neither wanted to risk losing. This was of course characteristic of any criminal trial— any time there were questions of jail time, reputation, money. But the prospect of losing took on an added edge for each side: for Linton and the county because some people—including colleagues in his own office—couldn't believe they'd ever wasted time and money with this unprecedented case; and for Reggie's side because he'd tenaciously denied any wrongdoing. Were he to lose, it felt like it would be shame piled on shame in the community.

With all this at stake, could there be a practical way out that let each side claim some victory? Just about the time of the April hearing, the county attorney, George Daines, reached out to Bunderson to talk about the possibility of a plea bargain. Again, this wasn't at all uncharacteristic of criminal cases.

There was one new bit of evidence that did give Linton and Terryl reason to dig in their heels. Four days before the hearing, they'd received the results from a study done by a forensic sciences firm in Salt Lake City that had asked the question: Was it likely that Reggie had hydroplaned?

The company investigator, Scott Kimbrough, concluded that, no, Reggie had not hydroplaned. "It is extremely unlikely that Reggie Shaw's vehicle hydroplaned. As his vehicle drifted across the center of the road, his wheels would have traveled over roughly textured asphalt that sat above the slight depressions of the wheel paths in the center of the travel land and would have been well drained," the memo from the investigator read in its concluding section. It noted that there was only "scattered rain" at the time of the accident.

The "dynamics of the vehicles involved in the accident strongly indicate that the vehicles had much higher levels of traction than would be available in a hydroplaning situation."

THE DAY AFTER THE April 15 hearing, Bunderson wrote a memo following a long conversation with a local lay official in the Mormon Church, Rod Merrell, who was the president of the Shaws' region. The conversation had to do with whether Reggie could complete his mission, still his stated dream.

Merrell referred Bunderson to Terry Johnson, who worked in the field for the Canada-Winnipeg mission. Bunderson called Johnson, who told the lawyer that this was the sort of case that would need to be reviewed at a high level of the Church, the Counsel of the Seventy or even the Counsel of the Twelve, otherwise known as the Apostles.

"There are no real hard and fast rules," Bunderson wrote in his internal memo. Reggie's ability to start a new mission, would be "a function of the type of crime, so I believe that if he is convicted of a lesser offense that would be helpful."

REGGIE LEFT FOR SALT Lake City. He lived in a spare bedroom with the family of a friend he'd made during his year in college in Virginia. Bunderson sent a letter on June 16 saying that he'd had several conversations with George Daines, "for the purposes of discussing a possible resolution in the case."

Reggie worked for the Bees and took classes, thinking he might study English. He met a young lady named Elise and they chatted on Facebook and hung out a bit. He met another young lady named Nikki.

Jackie, meanwhile, got closer with Gary, who came for that week-long visit in June. Leila made salads and read and watched television. For the Shaws, the legal bills crept up.

IN JUNE, JUDGE WILLMORE ruled on several of the motions that were discussed in the April 15 hearing. On the question of who would pay for Reggie's expert witnesses, Judge Willmore ruled that the state and Reggie could split the cost.

On the much more significant question about whether Trooper Rindlisbacher could be allowed to testify about his conversation with Reggie, the judge found the testimony could be allowed. He ruled that Rindlisbacher didn't have Reggie in custody and wasn't interrogating the young man in a way that would require a Miranda warning.

He wrote that the conversation was friendly and that Reggie had gone voluntarily into Rindlisbacher's car. "Defendant was not frisked or handcuffed and the tone of the conversation was cordial," Judge Willmore wrote. "No evidence was presented to the court that indicates Defendant felt like he was not free to leave."

This was a big one in the favor of Linton and the prosecution. If Rindlisbacher was free to testify then Bunderson would face a tough call as to whether to put Reggie on the stand, a proposition the defense attorney thought was fraught with peril.

At the time, the judge didn't issue a ruling about the other big issue, about Dr. Strayer's testimony. That would require deeper thought, and another hearing.

IN AUGUST, BUNDERSON WROTE Reggie for the first time with specifics of a possible plea, or, rather, of the offer from the county. It entailed Reggie pleading to two counts of negligent homicide, both class A misdemeanors. Reggie would be required to serve thirty to ninety days in jail, "watch a video of one of the deceased person's lives," and "for a period of six months they want to use you as a poster boy" for deterrence on texting and driving.

There was one thing Bunderson liked about the deal: It would be a plea in abeyance, a term of art that means the charges would be dismissed, removed from the record, if Reggie met the conditions of the sentence.

Still, Bunderson concluded, "I don't expect you to find this what you want, but we do need to consider it as a settlement and discuss it thoroughly."

ON SEPTEMBER 23, 2008, in Florida, half a country away, another tragedy occurred: A driver of a semi carrying a delivery to Home Depot slammed into the back of a school bus that had stopped to let children off. The bus had its warning lights on and its signs out, alerting drivers it had stopped.

The crumpled bus started to burn. Trapped inside was thirteen-year-old Frances Margay Schee. She died in the flames.

The semi driver said he never saw the bus. He told investigators he had been on the phone.

IN THE FALL, THE season ended for the Salt Lake Bees. Reggie got a job with the National Basketball Association's Utah Jazz, taking drink orders in a fancy club at the arena.

Then, in September and October, there were more developments on the legal and personal front that underscored the numbing effect, the purgatory.

On September 29, Bunderson wrote Daines to put in writing the proposed settlement that had been the product of ongoing conversations. It was much as Bunderson had described it to Reggie in August, but there was an important caveat to the plea in abeyance. They would get Judge Willmore to agree to "hold the guilty plea in a confidential file so it doesn't get recorded anywhere." This was aimed at maintaining Reggie's reputation and paving his path to return on a mission. It was intended to draw as little attention as possible to what happened on the morning of September 22, 2006.

This more serious discussion of settlement reflected the reality that an actual trial was quickly approaching. It was only prudent to take these steps. It also reflected, though, that Reggie's will seemed to be wilting, something that was tough to put a finger on—he never said "I did it"—but the fight seemed to be going out of him bit by bit.

THINGS WITH NIKKI DIDN'T work out. Reggie couldn't really open himself up. He wanted to, wanted to connect, but he was holding so much back. He talked about the accident in a distant way. Then there was Elise, the woman he'd met in the summer. She was blond and Reggie found her very attractive. They'd stayed in touch on Facebook. He asked her out again in October.

For their second date, he planned to take her to Bonwood Bowl, a bowling alley with a lounge and café on South Main Street in Salt Lake City, and then to a drive-in movie. He went to Elise's house to pick her up.

While she finished getting ready, he sat down on her bed and looked around. He saw something taped to the side of her mirror that caught his attention, a newspaper article. He walked over and read it. It was about him, how he was facing charges for killing Jim and Keith, and was scheduled to go to trial. Elise walked back into the room.

"Where did you get this article?" Reggie asked.

"My dad gave it to me. He's concerned about my driving behaviors."

"Have you read it?" Reggie asked.

"Yeah."

He smoothed back the article. "Do you know who this is?"

"No." She couldn't figure out what he was getting at.

"I want you to come over and look at this."

Elise walked over. She read the first couple of lines. She started to cry. She put her arms around Reggie.

Later, he thought, it was a funny moment, but an awkward one. Something he should have laughed about. They went bowling and to the drive-in movie. He had fun. They connected. He called her to go out again. But he didn't hear back.

CHAPTER 35

HUNT FOR JUSTICE

ALL THE STRANDS OF this story—the accident, attention science, the law, and Reggie Shaw—came together on December 11, 2008. It was a two-hour-and-twenty-three-minute court hearing, on its face just one more procedural step in more than two years of dug-in fighting. In this seminal case on texting and driving, the hearing examined how decades' worth of attention science might be applied to the law. And, crucially, it showed in the fullest detail yet just what really may have been happening inside Reggie's mind on the morning of September 22, 2006.

Reggie, already worn down and terrified of trial, was left transformed.

"AT THIS TIME, I'D like to call Dr. Strayer to the stand," Linton said.

David Strayer walked to the witness box. He wore a tan suit over a royal blue shirt and striped tie. His hair, a brown-and-gray mash-up, was shortish and scruffy in a way that might seem hip on someone half his age but gave him the look of an affable academic.

He took an oath promising to tell the truth. At Linton's request, he talked about his education and training, his voice a decided tenor.

"My area of expertise is attention and performance," he said. When he spoke, he sometimes looked up and to the side through his glasses, as if searching for what he wanted to say. He clasped and unclasped his hands when making a point of emphasis. About 30 percent of his research over the years had focused on driving issues, but in the more recent years, it had been the bulk of his research.

Dr. Strayer said he'd published more than fifty articles in major scientific journals, around twenty of them on issues pertaining to driving impairment, attention and distraction behind the wheel.

Reggie, sitting in his usual spot next to Bunderson, on the left side of the room (from Judge Willmore's perspective), looked impassive as usual. He chewed gum. And yet he was primed with nerves. The trial was coming up. That could mean a felony, jail time, real stakes. On top of it, he just couldn't stand glancing up at the families of Jim and Keith.

They were sitting just behind Reggie's parents. In between Jackie and Leila sat Terryl. She was concerned about how to prepare Jackie and Leila for trial. It was Terryl's impression that neither woman wanted to testify, to relive the thing. And they both just seemed incredulous that, with all this evidence, Reggie would contest the charges and there would have to be a trial at all.

Soon, though, Terryl was riveted by Dr. Strayer. All along, Linton had told her that they needed an expert, and they'd found *the* expert. She was hoping this testimony was going to seal it, vindicate all the effort, the pretrial preparation, the way she and others had extended themselves to take on Reggie.

As Dr. Strayer began, Bunderson, sitting next to Reggie, had a powerful, initial visceral reaction: This Dr. Strayer sounded very credible, like a good witness. All the more reason to pick apart the testimony and get as little of it admitted as possible.

THE PURPOSE OF THE proceedings was to determine to what extent Dr. Strayer's testimony would be allowed at trial. Was it relevant and based on sufficient and credible information?

At Linton's request, Dr. Strayer described the driving simulator he used to test how well people drove. The simulator was in a small windowless room down the hall from his office at the University of Utah and was basically the seat and dash of a Crown Victoria. On a computer screen in place of the windshield, the study subject would see animated views of cities and highways and navigate among them while, say, trying to text, or while dialing a phone, or talking on a phone, or, for the sake of comparison, while lightly intoxicated (under, obviously, a tightly controlled laboratory setting).

Dr. Strayer talked about other equipment, including the EEG, which allowed the researchers to "measure brain activity," and "various changes in brain activity as someone is engaged in an activity—driving or texting or talking to a passenger." They used eye-tracking technology to study where a driver was looking, say, at the road or at the phone.

Linton asked what Dr. Strayer had found when it came to texting and driving.

"The scientific data says there is a sixfold increase in crash risk."

That's because, he explained, the distracted driver didn't watch the road, lost track of the lanes, might miss hazards that came up or, even when seeing the hazards, lacked the focus to be able to react. He said that when drivers text they lose "just about all characteristics associated with safe driving."

He repeated that the texting driver faces a sixfold crash risk, whereas a driver talking on the phone faced a four-times increase in likelihood of a crash, which he said was roughly equivalent to someone who is legally drunk. A drunk driver and a person on a phone were equally likely to crash, whereas "we're seeing the risk factor for accidents when someone is texting exceeds the level when people are legally drunk."

"How many times have you testified in court?" Linton asked, pacing the middle of the courtroom floor.

"In terms of text messaging, this is the first," Dr. Strayer answered.

But, he said, in about thirty other cases, he'd been deposed—meaning interviewed by attorneys in a case but not in the actual trial—on the subject of the general risks of using a phone while driving. And he'd testified once, he said, in an actual trial.

"All of those are civil cases?" Judge Willmore asked.

"Yes."

"No criminal cases?"

"I've never testified in a criminal case," Dr. Strayer answered the judge.

"To my knowledge this is the first criminal case I've found on this issue." Linton directed this comment to the judge.

"You'll be able to get into that later."

It was Bunderson's turn.

THE DEFENSE ATTORNEY STOOD.

"You couldn't testify whether texting caused this particular accident?" he asked Dr. Strayer.

"Not this particular accident."

A confusing exchange ensued that is nonetheless significant. It had to do with the use of the phrase "significantly greater risk." Bunderson began by asking Dr. Strayer whether a "significantly greater risk," which is a relative term, implied that it was a "substantial risk," which is a more objective term. In other words: Is the fact that texting creates *more* of a risk the same thing as saying that it is an inherently substantial risk?

There are plenty of examples where a behavior is an increased risk without being particularly risky. For instance, eating in a car might be more likely to cause a crash than not eating. But that doesn't necessarily mean that eating in a car is a major risk that should be illegal. Or, what if a regular coffee drinker drives to work in the morning without his usual cup of caffeine? He'll be less alert on the road, but few people would say he shouldn't be allowed behind the wheel.

Now, it might be completely true that the risks of texting and driving are inherently risky. But that was, Bunderson was arguing, beside the point under Utah law. He was arguing that Dr. Strayer's testimony would run afoul of a rule preventing him from suggesting a legal conclusion to the jury.

Linton was beside himself. What Dr. Strayer was saying, he argued in a back-and-forth that followed, was merely fact. It was a fact, he said, that there was a six-times increase in the risk of a crash. That wasn't inference. He objected to the line of questioning.

Judge Willmore didn't see it as so open-and-shut. As he looked at the Utah statute while sitting on the bench, he appeared almost pained. He was trying to figure out whether Dr. Strayer might overly influence the jury were he to testify to the inherent substantial or significant risk.

"I'm overruling the objection," he said.

The defense attorney turned back to the academic. "When talking about driving, I remember that old song: 'Keep your hands on the wheel and eyes on the road ahead . . .'"

Dr. Strayer interjected. "What we've found even more on then those two is your mind on the road," he explained. "Something that takes your mind off the road, or hands off the wheel, creates a risk. But [the bigger risks comes from] . . . mind off the road than hands off the wheel."

It was a powerful point, made in an offhand way, but it hit Reggie hard. Bunderson seemed to want no part of it. He redirected the conversation.

"As far as texting, have you actually published an article on texting?"

"No, what you've seen is a draft version," Dr. Strayer answered, referring to the paper showing a sixfold increase in crash risk.

Bunderson asked if anyone had published a paper on the risks of texting.

"Yes."

"When?"

Dr. Strayer said he couldn't recall exactly, but it was sometime between 2005 and 2007, probably 2007.

Bunderson asked where it was published.

By researchers at a university in Australia, Dr. Strayer answered.

Bunderson asked if it was available in the United States and Dr. Strayer said everywhere.

"Are you aware if it's been published in any way that became known to the general public?"

"That I don't know."

WHATEVER THE LEGAL IMPLICATIONS of Strayer's testimony, it was scoring big points with Reggie. He found it revelatory. What he was hearing, and beginning to comprehend, was that it was possible for him to have his hands on the wheel and eyes on the road, and still be lost somewhere else. He thought: *I was so distracted by my phone device that I wasn't aware what was happening in the car, and had no memory of it.*

That dawning idea got reinforced just a few moments later, when Bunderson asked: "Did you ever in your study have someone text and then quit texting and determine how long it takes to get back into safe driving mode once they've quit?"

"We have been looking into that. It's not easy to determine how long impairments last—generally those costs can persist for ten seconds or fifteen seconds."

Even after the texting ends? Bunderson asked.

Yes, Dr. Strayer explained, noting that it takes time for the person to become reacquainted with the road. "Depending on the complexity of the driving task, it may take fifteen seconds or more after you've pushed 'send' before you're fully back in an unimpaired state."

From the perspective of the law, this was another moment where Dr. Strayer was touching on the state of mind of the person texting. Judge Willmore interrupted and asked the professor to elaborate. Dr. Strayer said he was presently doing experiments in how long it takes to regain "situational awareness" once a person was finished texting.

In his seat, Reggie continued to look distant, but increasingly he was anything but. The words were reaching him, he later reflected, in a way that nothing else had in the preceding twenty-six months.

Dr. Strayer said: "After you've pushed 'send,' and you're waiting for a response, fifteen or twenty seconds may pass before you're fully back and have regained your sense of all vehicles around you."

An exchange followed with Bunderson, who bored in on whether this could be proved and what was the basis of the proof, especially considering that Dr. Strayer had not completed his study. Dr. Strayer said that there was a long history of research into the concept of switching costs, the time it takes to refocus after moving from one mental task to the next. This, he emphasized, is not a controversial topic.

In effect, he was saying that when someone is finished texting, their minds can still be so absorbed in the task that they do not know what is going on around them. To reinforce the point, Dr. Strayer described the concept of "inattention blindness," in which lack of attention can cause someone to not see what's going on around them.

Pointedly, Reggie thought: *What if I was so preoccupied that I actually didn't know what was going on?*

Looking back, Reggie says:

> *When he talked about inattention blindness and how someone can get a text and read it, or send a text, and for about fifteen seconds afterwards they can drive down the road and not pay attention and not be focused and still be mentally focused on the phone, that's when the thought occurred to me, like: "Hey, it's clear you were texting while driving. You did it all the time, the phone records show it. It is obvious, there's no debate."*

Reggie was, as if co-opted by a text while driving, preoccupied by revelation. Hearing him retell it—seeing how impassive he looked in a videotape of the court proceeding, it's hard to make sense of how singular and important the moment was for him. Then again, it was coming in this larger context of the prospect of a trial and jail, and of the months of it all eating away at him. The accident was already defining him in big and little ways, like on the date with Elise, the girl who had the article about him taped to her mirror.

If Reggie was looking for a way to explain to himself how he could've lied, or deceived himself, or how he might just not have grasped what was going on at the time of the accident, Dr. Strayer gave it to him.

IN THE COURTROOM, BUNDERSON kept hammering away. He asked Dr. Strayer if talking on a phone while driving was against the law in Utah.

No, Dr. Strayer said.

What about texting?

No, Dr. Strayer said, except that he believed it was illegal for a period of time just after someone got his or her license, new teen drivers. Dr. Strayer said that, in a few states, like California, talking on a phone was illegal while driving. (In California, drivers could talk using a hands-free device.)

Bunderson asked whether there's a general knowledge of the risks of using the phone while driving, given years of research.

Yes, Dr. Strayer thought there was.

"Is it fair to say that, as compared to talking on a cell phone or driving under the influence, the general public hasn't had information about the risks of texting and driving?"

"In terms of the scientific literature, you are correct."

Bunderson was pinning down the idea that the general public might not be aware of the risks of texting and driving or, if they had some passing understanding, it wasn't necessarily born of a mass of scientific data. So how could Reggie have been expected to know twenty-six months earlier?

And Bunderson had given Judge Willmore heavy pause about the use of the term "significantly greater risk," which might imply by its definition "substantial risk." And, already, several times, Judge Willmore had stated he would allow no comparisons to drunk driving because, the judge said, they would be too "prejudicial" for the jury, meaning they would draw a comparison that carried too much negative weight.

Were this a prizefight, Bunderson was scoring a lot of punches.

It didn't matter. Bunderson didn't know that the case was already over. Reggie, still showing nothing on his face, could no longer deny it: He, his texting, had killed Jim Furfaro and Keith O'Dell.

THE TESTIMONY AND QUESTIONS wore on and the hearing wound down. More details, more probing by Bunderson. Finally, they concluded.

Linton made a general plea, asking Judge Willmore to recognize that this was an important moment in history.

"This is a new technology," he said. He compared the lack of awareness of the impact of electronic gadgets on drivers to a time "where police officers would catch someone with a DUI and put them back in their car and tell them to drive home.

"The new tech is growing at exponential rates and we're asking the court to take that into account."

Judge Willmore said he was prepared to offer a preliminary conclusion: He would allow some of Dr. Strayer's testimony. But it would be limited. "He cannot testify with regard to the mental condition of the defendant in a criminal case."

There could be no mention of DUI.

As to the sixfold increase in crash risk caused by texting: "That's the one I'm struggling with more than anything.

"I'm really struggling, Mr. Linton," he told the prosecutor, and asked Linton to make his case again about why it should be allowed.

"It's a fact," Linton said. "It's a fact there's a six-times greater chance of being involved in an accident," he continued. It's up to the jury to decide if Reggie should've known that, if it was criminally negligent to not be aware of that.

By comparison, Linton said, a ballistics expert could testify that someone had shot another person, but that doesn't mean the shooter "shot that person on purpose."

He looked tired and strained when he concluded: "We cannot look inside of a human brain."

Judge Willmore took it in. "These are tough issues. Tough cases make hard issues."

Then he summed up. "We're set for trial on February eighteenth, nineteenth, and twentieth," he said. There would be a final pretrial hearing on February 2. "If there's going to be a plea bargain, it has to be by that date."

THERE WOULD BE NO trial.

Reggie was ready to deal.

Watching him from the spartan gallery, Terryl didn't pick that up at all. Nothing on Reggie's face suggested a change in his position. Besides, Terryl was having a different reaction to Dr. Strayer's testimony. It was, if not elation, then something close to it. It was a feeling that things were finally, decidedly, breaking toward the prosecution. Even though Bunderson had raised some serious doubt, Terryl felt the core science was so powerful as to overwhelm any doubts.

"This is it," she recalls thinking at that moment. "We've got him. We've got it nailed."

Reggie and Terryl had seen each other plenty of times. But they'd never met. The hearing with Dr. Strayer made the meeting imminent. Reggie was getting ready to cop to what he'd done, but Terryl, as she and others put it, was "out for blood."

PART THREE

REDEMPTION

CHAPTER 36

HUNT FOR JUSTICE

O N MAY 3, 1980, thirteen-year-old Cari Lightner was walking to a church carnival in Fair Oaks, California, when she was run down by a repeat-offender drunk driver. Cari's mother, Candy Lightner, vowed to take on drunk driving and founded Mothers Against Drunk Drivers, later renamed Mothers Against Drunk Driving.

It became, arguably, the most powerful, and effective public safety advocacy group in the country. It led the charge for tougher laws and tougher enforcement, and for public education. In 1982, alcohol impaired drivers—drunk drivers—caused at least 21,113 fatalities (some estimates have that figure as high as 30,000 deaths in 1980). By 1991, that figured dropped to 15,827. And by 2010, the figure had fallen to 10,228, a drop of 52 percent, according to the National Highway Traffic Safety Administration, a division of the U.S. Department of Transportation.

Alcohol-related deaths still constitute about one-third of all traffic fatalities, but the change in culture and behavior has been profound. After all, prior to MADD, it was considered socially acceptable to have one last drink at the party before getting on the road.

So, too, there has been profound improvement in the use of seat belts.

Today, seat belts are worn close to 90 percent of the time. That compares to about 15 percent in 1983. The monumental change owes to a marked shift in public understanding that seat belt use saves lives—it's the single most valuable thing someone can do to spare themselves death or injury in the event of a crash—says Barbara Harsha, who for twenty-five years was the executive director of the Governors Highway Safety Association, a group made up of state traffic-safety experts. (Harsha retired in 2013.)

In 2008, as the Reggie saga wore on, Harsha and others were trying to apply what they knew about drunk driving and seat belts to the issue of cell phone distraction by motorists. There were some caveats to the comparisons. Notably, the effects of wearing seat belts and drunk driving had been tracked for a long time and their impacts were clearly measurable, but the data set on distracted driving was just not there. Moreover, a drunk driver remains impaired for the whole ride, whereas the impairment of a cell phone user is intermittent, and impacts the brain only in the period surrounding their interaction with the device. So safety advocates cautioned against making a perfect comparison between driver distraction by cell phone and drunk driving.

Another difference, according to safety expert Bill Windsor, is that no one ever told people that drinking and driving was a wise idea, yet there has been a widespread idea in the culture that it is good to be digitally connected all the time. "There is a lot of pressure on kids and even adults to stay connected," says Windsor, who is the assistant vice president of consumer safety at Nationwide Insurance and sits on the board of directors for MADD and the National Safety Council.

Still, in broad terms, the safety advocates hoped the successes with seat belts and drunk driving would provide a blueprint for addressing cell phone distraction. In a nutshell, those accomplishments owed to the combination of two principles: enforcement of tough laws, plus heavy public education (of the dangers, the legal costs, and human toll). Enforcement and education "are proven to work," Harsha says.

Of course, there had to be laws on the books before one could press to enforce them. In 2006, at the time of the accident that took Keith's and Jim's lives, there were no laws explicitly covering texting and driving, except several prohibiting the behavior among novice drivers. Two years later, a handful of states had succeeded in banning texting or use of handheld phones by all drivers.

Enforcement remained a serious problem even where there were laws, notes Harsha and other public safety advocates. "People text in their laps," Harsha says, which makes it tough for police to see what's happening.

Also, people were outright lying, which was driven either by self-deception or survival. "It's fairly rare that somebody is going to admit they were talking on a cell phone," says Mr. Windsor. As Reggie's case showed, it was not easy to get the cell phone records and proof.

And so all these roadblocks were making it hard to even measure who was doing the behavior, in a way that, say, Breathalyzers or field sobriety tests could measure for drunken behavior. How many people were on the phone when they got into a wreck? Most police agencies still weren't collecting that data, or even being required to ask. There were a lot of horrific anecdotes, and there was some powerful implicit evidence: Traffic deaths remained at epidemic levels despite all the other improvements to safety, not just the fall in drunk driving fatalities and the huge increase in the use of seat belts but also in the billions spent on other safety measures, like air bags, safer roads, antilock brakes. To Dr. Strayer and others, the answer is that the rise in driver distraction, caused in large part by new technology, has undermined the gains those other safety measures have made.

Others interpreted this lack of clear evidence differently. Cell phone industry officials said that the fact traffic deaths hadn't soared meant that distraction from electronic gadgets was being overblown as a threat.

At the time, texting was still relatively new, but talking on the phone and driving was by now ingrained and done widely. In 2008, the federal government published a paper that found that during 2007, 11 percent of drivers at any given time—1.8 million drivers—were using a phone during daylight hours.

Even as the behaviors were becoming ingrained, many drivers thought it was dangerous and stupid. The AAA Foundation for Traffic Safety, an arm of AAA (formerly, the American Automobile Association), did a survey in 2008 in which drivers said that the two biggest things that could be done to "prevent serious motor vehicle accidents" were: reduce driver distractions and reduce or avoid cell phone use. Those two ranked above asking drivers to slow down and reducing drinking and driving.

Elsewhere in the survey, drivers ranked drunk drivers as the first most serious traffic safety problem, and cell phone–using drivers as the second.

In other words, cell phone use by drivers was ranking as a serious problem by the public, and, yet, we were doing it all the time. "The culture is: 'It's not me, it's you. I'm the good driver,'" Harsha says.

The safety advocates were confounded. Something was not quite adding up, this disconnect between attitudes and behaviors.

"We don't really know what is going to work. We know it's going to take a really long time."

There was one more thing, though, that the safety advocates felt they'd learned from MADD. Something visceral and less tangential than, say, statistics, laws, police enforcement, or television commercials: The issue had to be made personal.

Says Harsha: "One of the lessons of MADD is you've got to put a face on the problem."

SOMETIMES, WHEN DR. GAZZALEY thinks a lot about a problem, the answer comes in the form of a dream. He awoke to such a dream one morning in late 2008. In it, there was an idea for an experiment. It was counterintuitive: Could a video game—the very technology that often seems to distract someone by drawing them away from other priorities or to shorten attention span through habituating players to constant stimulation—hold the keys to training people to sustain attention?

He contacted some people he knew from LucasArts, a big-time video game studio founded by *Star Wars* creator George Lucas. He told them he wanted to marry technology and brain sciences to create a scientifically engineered therapy for waning attention. In his vision, he'd prove that the game worked by using imaging technology to show improvements inside the brain.

He put together a grant proposal. They'd need a lot of resources, and it didn't seem like it would be easy to get people to pony up. They'd need to challenge a lot of conventional wisdom.

AND DR. STRAYER MADE a discovery that seemed just too weird. It happened unexpectedly during an experiment he had run in his driving simulator. The idea was a twist on his distracted driving studies. This one explored whether the task of driving actually interfered with the task of talking on the phone. How accurate could the driver be at thinking through complex issues, even doing math problems presented by the person on the other end of the line? In this case, the primary task was considered to be the talking—that was the top-down goal—while the driving was secondary.

After the subjects went through the simulator, a senior thesis student plotted the individual performances on a graph. As expected, the talking performance declined significantly. Except for one study subject. There was this dot on the graph that didn't make any sense.

One person seemed to have shown slight improvement, coming across on the graph as an anomaly, an expert multitasker.

"This has got to be screwed up somehow," Dr. Strayer thought. "This doesn't fit with anything we know."

The researchers looked at the coding of the data. Maybe they had made a mistake. "We tried to make it go away," Dr. Strayer says. They spent a week looking at the data, but the outlier remained every time.

"Let's go see if we can find more of these people."

ONE SPRING DAY, DR. Greenfield—the addict turned psychologist and addiction specialist—got a new patient in his Connecticut office. It was a sixteen-year-old boy who was spending six hours per day on the Internet playing online games. His parents were beside themselves, unable to get him off the computer. They forced him into counseling. He was defiant. His grades were decent. What, he wanted to know, was the problem?

It was a basic power struggle and Dr. Greenfield adopted what he describes as a "paradoxical approach." In his early meetings with the young man, he took the boy's side. "It may not be a problem. It's just a game, after all. Games are fun," he told the patient. "The problem may be what a bunch of people think about what you're doing." Dr. Greenfield urged the boy to counter the parental arguments, to start looking at the situation from different angles.

At one point, the boy revealed that perhaps he *was* spending a lot of time online. "If you want to play a little less, maybe I can help you do that, if that's something you want to do," Dr. Greenfield responded.

As Reggie's saga was leading to an unexpected end, the people on the leading edge of public policy, neuroscience, and addiction sciences were grappling with, and discovering, how to rein in the darker sides of technology use. Could it be tamed? Were some people more able than others? Could technology itself hold the key to helping us remain master over these powerful machines?

CHAPTER 37

REGGIE

THE MEETING TOOK PLACE, like the others, in Bunderson's conference room in his firm's small pink-hued offices. Mary Jane and Ed joined Reggie, of course. It was January of 2009. Bunderson discussed the upcoming trial, just days away, and explained the latest plea agreement. Reggie could take a sentence that ranged up to ninety days in jail and included community service. And were there to be any legislative hearings about texting, Reggie would have to testify. Such an agreement, called a plea in abeyance, would allow for his record to be wiped clean, should he meet all the conditions.

Bunderson said he thought he had a winnable case, but in trial there was no sure thing.

All eyes turned to Reggie.

He nodded. He didn't say it, but ever since Dr. Strayer's testimony, he was obsessed by the idea that he had been texting, maybe not at the precise moment of the crash, maybe a moment before, he couldn't exactly remember. It was blurry, as he'd always said. But now that he'd let the idea in, he couldn't get around it: *I texted and drove and I killed those "two great men."*

After Reggie nodded his agreement to the plea, there was no argument. Ed and Mary Jane, as much as they feared their son going to jail, felt a deep weariness with the whole process. It had to end, somehow. Plus, crucially, they saw that Reggie had complete conviction.

Even though Reggie was coming to terms with his actions at the time of the accident, there was plenty of reason to think the plea was just expediency. At least from the outside, it seemed he was grasping for an option that avoided the risk of a longer sentence, and the lifelong stigma of conviction. Put another way, Reggie's team saw no reason to believe this was anything but a practical deal. They didn't sense yet, nor perhaps did Reggie, that it masked a true, remarkable transformation.

IT WAS COLD, PREDICTABLY, on January 27. Reggie Shaw's plea hearing was set for 3:30 p.m. His was the last case on the docket, and by the time they got to him, it was dark outside.

The hearing was short and completely anticlimactic. Reggie, legs shaking, accepted a plea in abeyance. He would get some jail time, no more than ninety days, to be determined by Judge Willmore. And he'd have some community service, which entailed doing a media interview, and talking to some schools. Here, too, the judge would work out the details.

In exchange, if the requirements were met, the thing would drop from Reggie's record. As if it hadn't happened.

Ordinarily in such a plea, there was not a formal sentencing. But this case was different, and Terryl and the others insisted that Leila, Megan, and Jackie have a forum to express themselves. Before then, Reggie would meet with a probation officer who would help suggest a sentence to the judge.

After years of slow back-and-forth, the families were stunned by the abrupt finality that confronted them—the hearing had taken mere minutes. Numb, they were hustled out of the courthouse, where there waited a few members of the media.

"We stood outside in the cold and dark. It had been months and months. They were going to plead and then they weren't," Leila reflects. "We were ready for a trial."

The whole thing concluded with a whimper.

"We really thought it was the end at that point."

CHAPTER 38

HUNT FOR JUSTICE

NOT LONG AFTER LINTON filed charges the previous summer, Terryl called Kaylene Yonk. She had long been a probation and parole officer for the Utah Department of Corrections. Now she had her own private firm where her job was to write presentencing reports, an official counsel for the judge about where a sentence should fall and why.

The prosecutor's office loved Kaylene, whom they saw as fair but tough. She saw herself as tough, and so did defendants. She stood six foot one, had played basketball and volleyball in college, and, as a sworn police officer, carried a Glock semiautomatic pistol. In past years, she'd carried a .357 snub-nose revolver, a compact yet powerful little handgun that fits nicely in the palm.

Every once in a while, when her four kids were little, she'd take one of them to work. While she interviewed a prisoner at the jail, she'd leave one of her kids in the holding cell and tell them, "This is what it's like" if you do something wrong.

Dealing with criminals and high stakes took its toll on Kaylene. She slept terribly. She got sinus headaches that turned to migraines. Daily, she took OxyContin, a powerful opiate, to cope with the headaches.

Terryl called her in early 2009 about Reggie and said he was a remorseless, arrogant liar. "I want him hammered bad."

She wanted the maximum end of the plea bargain. She'd hoped it would be more. Judge Willmore would have to approve it anyway, so Kaylene's recommendation mattered.

"Terryl is out for blood. She thinks this kid should go to prison for the rest of his life," Kaylene reflects.

KAYLENE'S SMALL OFFICE HAD room for her desk, two guest chairs, and a refrigerator in the corner that she stocked with Diet Cokes. On the wall was a framed embroidery of a Japanese geisha that Kaylene had made. There was a picture of a sailboat in the sunset, painted by her brother-in-law.

When Reggie came in on the afternoon of January 27, just after he accepted the plea, Kaylene's first impression was he was either arrogant, or very quiet, but probably the former. She didn't have his file so she started asking him a bunch of questions about why he was there, and his background. He answered in monotone, with a "Yes" or a "No."

Reggie said he didn't know if texting had actually happened right before or during the wreck. He said he texted all the time and that it was a nonstop kind of thing in his life.

Her opinion of him hardened. He was arrogant, and maybe not only that: Perhaps he had no emotion whatsoever. That troubled Kaylene and reminded her of just a handful of the thousands of convicts she'd worked with, including one woman from Smithfield who had chained a girl in her basement and shown zero remorse.

Reggie "came across like this was a business meeting."

After Reggie left, Kaylene was in agreement with Terryl. Maximum sentence.

THE SECOND MEETING TOOK place a week later. This time, Kaylene had the police report, including the pictures from the accident scene. She arrived less certain about what she was dealing with. In the intervening week, she found herself having trouble making sense of the idea that a kid who'd never done anything wrong, not so much as a speeding ticket, had been responsible for the deaths of two men.

She had the pictures from the accident in a loose-leaf binder. (The images were so disturbing that, at one later point, when she showed the pictures to her own son, he said, "This is the most awful thing I've seen in my whole life.")

When Reggie came in, she pulled out the photographs.

"I'm a firm believer that you need to see exactly what you caused," she said.

The blood drained from Reggie's face.

"I can't handle this. Please don't show me these pictures."

He started crying, and he couldn't stop. She put away the pictures, which she says was a very rare kind of restraint. "I didn't know how far he'd lose it, and I didn't know if I was ready to deal with someone who would lose it so totally that I couldn't handle it."

"He finally did break," Kaylene says. "All of a sudden, he turned into this nice little kid."

KAYLENE MIGHT NOT HAVE realized the extent of it, but it was a huge moment for Reggie, a floodgate. Reggie had let the accident in, perhaps for the first time in public. He'd done all his crying alone, not even with his family. Now it was starting to come out. His private and public selves beginning to reconcile. He'd begun to bridge the vertiginous gap between what he told the world and the truth he knew deep inside.

Kaylene listened to his story, and his explanation for why he'd never apologized to Keith's and Jim's families. Despite feeling desperate to reach out, Reggie had been instructed by his lawyer that doing so could be an admission of guilt. He told her that he'd had no idea that texting and driving was wrong; no one had told him. He said, "If it could've been me who died in that accident, instead of them, take me."

It was revelatory for Kaylene, too. Like others in this case, she was seeing Reggie through the lens of her life, and her own road-time behavior. "I was pulled over three times for DUI," she says. But she hadn't been drinking; it only looked that way. "I wasn't drunk driving. I was asleep." That had happened years earlier when she first started at the Department of Corrections and was working the graveyard shift in Ogden. In the mornings, she'd drive home and find herself nodding off. She was exhausted, stressed, and dealing with the headaches.

She also said she had some empathy in general because of her own medical challenges, the idea that she was just as human as anyone else. She got these terrible headaches. (They resulted, she says, from a series of surgeries she'd had earlier in her life to relieve severe sinus infections.) Then there was the daily forty milligrams of OxyContin she took to cope.

"I take narcotics every day and I drive on them every day," she admits, and she says she's done so for many years. She says she couldn't think of a time the drugs had impacted her driving, noting that she'd gotten used to a dose that, doctors say, could be manageable behind the wheel. She can, though, occasionally slur her words a bit and is prone to repeating herself.

"Have you ever fallen asleep at the wheel and had to shake yourself awake?" she asks rhetorically. "It's happened to me a lot."

To get in a wreck, she says: "All it takes is a half second of inattention."

At the second meeting with Reggie, she saw a broken person who she could connect to. A different Reggie revealed himself, one who virtually no one else in the process, at least outside of his family, had seen up to that point.

"I decided that putting Reggie in jail for one day or a year, it's not going to make any difference. As far as I was concerned at that point, we didn't need to send him to jail. He's already punished himself. It's eaten him alive and it will for the rest of his life."

Kaylene now had to have a direct, possibly tough talk about it with Terryl, who, remarkably enough, had just had her own harrowing driving accident.

THE FIRST TIME TERRYL had nearly gotten into a terrible wreck was when she was in Orange County. It was that awful day when Danny had snatched Mitchell away from her outside the movie theater. Driving home, distraught and reckless, speeding, she'd nearly gone off the side of the road.

There was nothing so excusable that caused her distraction this time—on a cold day in February. What distracted Terryl was a Twinkie. It was sitting in the front console, the last one, and Terryl had promised it to herself. Taylor and Jayme were in the back. It was snowing. They were heading for school. It was a busy area of Logan, often with joggers, but, on this day, quiet, owing to the weather.

Taylor started saying he was going to eat the last Twinkie. It was lighthearted son-mother banter. He reached for the Twinkie.

"It's the last Twinkie, and you don't get it," Terryl said, laughing, as she reached behind her.

She started to turn a corner. Now she was going too fast.

The car began sliding. She lost control. She spun. The car wound up backward in a ditch.

She had a cascade of thoughts. "I said: 'This could happen to any-body,'" she says of making a driving mistake. "If someone were out there jogging, I could've killed somebody."

But she didn't feel this let Reggie off the hook, because his action didn't have theoretical consequences but actual ones. And she still saw a distinction between her momentary lapse of judgment and focus, and what science said was a widespread and systemic problem caused by cell phone use.

A FEW DAYS LATER, Terryl and Kaylene talked. Terryl was miffed that Kaylene was thinking about suggesting to the judge a lesser sentence than she and the prosecutors wanted.

Terryl impressed upon Kaylene the idea that Reggie had swerved into the oncoming lane several times before the wreck. After the first swerve, he could have and should have realized the risks in his behavior and pulled over if he wanted to text more.

"There's no excuse," Terryl told her. "He should've pulled over before he was crossing the center line."

Kaylene asked Terryl: "Haven't you ever fallen asleep at the wheel, or dropped something and reached for it, or went over the center line? If you did, did you stop and pull over and stop doing whatever you were doing?"

Terryl confessed the "Twinkie incident" to Kaylene.

But she maintained that cell phone use was happening more and more, becoming a regular part of driving, and probably dangerous. And the consequences were entirely different, Terryl told Kaylene. That was the bottom line. Besides, she thought: *Spare me Reggie's proverbial deathbed conversions and the last-minute apologies, in the face of jail time. This guy has stonewalled and lied.*

And then there was a new affront.

Part of the plea agreement was that Reggie would give a media interview to Nadine Wimmer, the anchor for Channel 5 in Salt Lake City. But just a few days after Kaylene and Reggie met, Terryl got a call from the anchor.

"'I can't do this interview,'" Terryl says Nadine said.

"Why not?"

Nadine explained that Reggie's lawyer had called and said that Reggie could be interviewed but would only talk about the risks of texting and driving, not about his experience in the accident.

Terryl was irate. Reggie would skate with a little jail time and community service, and then have his record wiped clean. Nothing good would come from this. Modest justice, and little learned.

CHAPTER 39

THE LAWMAKERS

O N FRIDAY, THE THIRTEENTH of February 2009, at 2:15 p.m., after a long week of legislating, the Utah House of Representatives met to discuss texting and driving. Ten of the twelve members of the Law Enforcement and Criminal Justice Standing Committee convened in the capitol, the meeting called to order by their vice chairman, Curtis Oda, a Republican and insurance agent. They gathered in room 25, nothing fancy, more like city council chambers than a rotunda, the committee members in a horseshoe at the front, and a smattering of people in the two aisles of chairs looking on.

On the committee's agenda were two proposed pieces of legislation restricting wireless phone use by drivers. One was from Representative Stephen Clark, the contractor from Provo. It called for a ban on texting while driving. The other, proposed by another Republican who was a finance professional, was slightly more aggressive than Clark's in that it called for a ban on texting while driving as well as a requirement that drivers in a school zone use a hands-free device.

When it came time to take on the texting bills, Clark, friendly with a round face and round glasses, began by saying he would try to keep things brief "so you can move on it quickly." He was being polite, the day was late, but this was going to be a fight. In the preceding days, he'd been pulling members of the committee aside in the caucus room to lobby them.

"We need to do something about this," he told Representative Carl Wimmer, the policeman turned legislator. Wimmer recalls answering, "No, we don't need to *do something*. That is the mantra of big government."

Wimmer says that Clark asked him, "What will it take to make this work for you? What can I do to get your vote? What do I need to do to make this palatable?"

"My answer was: 'Nothing. Nothing. I will not run this bill,'" says Wimmer.

Generally speaking, Wimmer was considered among the more conservative legislators, but not a total outlier. For him, being conservative meant supporting extremely limited government and states' rights. To take an example, he wanted the state of Utah to withhold federal income tax paid by Utah residents until a state committee determined which federal programs were constitutional. Only then would it release the income tax money to the federal government to be spent on things it deemed constitutional (and not, he said, for unconstitutional things, like the Department of Education). On other issues, his view of states' rights was gummier. For instance, he liked the federal Defense of Marriage Act, defining marriage as being between a man and a woman, because he feared that, without it, some states would wind up pressured to allow gay marriage.

"I'm not one hundred percent consistent about everything," he concedes.

Gregarious and happy to engage on any topic, Wimmer jumped right in during the hearing to express his concern about the proposed texting ban. He read from a law passed just a year earlier that made it a class A misdemeanor to drive carelessly for any number of reasons, including using a mobile phone.

"We've got twenty or thirty laws on the books that make it illegal to drive unsafe—you cannot speed, cannot drive left of the yellow line and right of the white line. You cannot wreck. You cannot follow closely," he told his fellow legislators in a sum-up of his feelings on the subject. "One more [law] is not going to make a difference."

Several other legislators responded by arguing that the point was to turn the ban against texting into its own offense. That way, it would send the message that texting was wrong and, more to the point, police wouldn't have to wait until someone drove dangerously to pull them over.

"We would not sacrifice a child before we charge somebody," said Representative Paul Ray, a Republican who was the legislator who had proposed the second, stricter, piece of legislation banning texting while driving.

A back-and-forth followed among a handful of the legislators. Most notable was an exchange between Clark and John Dougall, a Republican and electrical engineer and businessman who would later become Utah's state auditor. He had short black hair and a question for Clark: Don't you trust our citizens to make good decisions behind the wheel?

"I guess I do not," Clark responded. Adding: "Like that guy this morning." It was a reference to a story he'd told earlier in the hearing about some guy he'd seen texting while driving erratically.

At another point, Dougall pointed out that people are allowed to drink a bit and get behind the wheel but they're not allowed to drink to the point of being drunk. What he was getting at was that people can be trusted to text a bit, to the point that they're not a danger.

Nearly an hour and a half had passed, it was almost five p.m. The outlines of the conversation were clear: up with freedom and down with big government. There was still time for a few public comments.

A WOMAN, WHO SAID she was the mother of six children and the grandmother to twenty-one, stood up. She described herself as just a citizen alarmed by the growing phenomenon of texting and driving. She said it wasn't just teens, but moms and grandmas, which she realized when she talked about it with a group at a Christmas party a couple months back. "Everyone is doing it."

The opposition came from two speakers. One was a representative of the Utah Association of Criminal Defense Lawyers. He wondered how police would be able to prove what someone was doing on their phone—texting, talking, emailing, what?—without being intrusive into the motorist's personal business.

The other comment, the one that seemed to sum it all up, came from an insurance agent named Michael Tingey. He echoed Wimmer's concern that this was just another law when there are plenty of laws on the books. But he really got a head of steam when discussing the comparison between drunk driving and texting while driving.

"The University of Utah study linking drunk driving or cell phone usage is patently ridiculous. It's inaccurate, and I can prove it to you right now."

The insurance man pointed out that far more people in Utah use cell phones than drive drunk, but challenged the committee members to think of how many fewer deaths they'd heard about from texting than drunk driving.

"I really reject and resent the fact that people are saying that cell phone usage and drunk driving are the same," he said. "To try to equate them is simply wrong and everyone knows it."

He said that what was happening was a good old-fashioned case of political correctness, the vilification of drivers and their cell phones. And he warned that "stampeding" to stem it could lead to unintended consequences. By way of example, he said that everyone talks about the number of children killed in school massacres, but said there had been fewer deaths from such shootings from 1996 to 2001 than there had been during that time caused by air bags deploying in cars and killing children.

"We didn't get the public outcry over air bags," he said. "Why? Because it's politically correct to point at guns and say this is a terrible thing."

Same thing, he said, with cell phones.

Dr. Strayer had been invited to speak to the legislature, but he was traveling. He'd testified in front of the legislature before, and he didn't figure his voice would make all that big a contribution. He put the odds that this committee would pass on the legislation for a vote by the full house at "zero."

And, on this day, he was right.

Shortly after Tingey, the insurance man, concluded, the committee took up a vote. There was a motion to pass Clark's bill out of committee, but it wasn't seconded. A motion was passed to move on to the next item of business.

This legislation seemed headed for the cutting-room floor.

CHAPTER 40

THE LAWMAKERS

DOUGLAS AAGARD WAS IN a hurry. The chair of the Law Enforcement and Criminal Justice Standing Committee was staring down seven agenda items on another Friday afternoon. Of the twelve members of the committee, five were elsewhere, flogging some or another piece of legislation. And Representative Clark, the sponsor of the texting ban, which he'd gotten on the agenda for a final airing, wasn't anywhere to be found.

So the Chair moved to the next agenda item. It was sponsored by Wimmer. It was a proposal to extend the statute of limitations on environmental crimes. "Believe it or not, this is a pro-environmental bill," Wimmer began, eliciting a chuckle. It passed unanimously out of committee.

Clark still hadn't returned. Someone suggested that, in his absence, Aagard allow a few more comments from the audience on the texting bill, H.B. 290, officially: The prohibition of Wireless Communication Device Use in a Motor Vehicle.

Aagard thought that made sense—the place was packed for this hearing—but in asking for comment, he offered a preface: "Bear in mind, we've pretty much heard this bill last time so unless you've got something to add . . ." In other words, we've seen how this plays out, so let's get it over with and move on to something that this committee intends to pass; let's quickly move past this last-ditch effort by Clark.

"We need you to identify who you are and who you're with."

A petite blond woman walked to the front. She had a piece of paper in her hand.

"My name is Terryl Warner. I'm with the Cache County Attorney's Office in Logan." She wore a black dress with a forest-green blazer over it.

"In July of 2008, a woman in Salt Lake City ran a red light while trying to send a text message. She critically injured one driver and killed a pedestrian," Terryl launched in. Her voice came across rapid-fire, bouncing, like a rabbit dashing across a warren. She recited some of the research she'd been compiling. "Weeks later, a train operator texting while on the job killed twenty-six people and injured nearly one hundred and fifty. In March of 2007, seventeen-year-old Lauren Mulkey was killed when a texting driver ran a red light. Several months later, five cheerleaders were celebrating after their graduation and they were killed when the driver was texting. In September of 2006, two rocket scientists were killed when a text-messaging driver crossed the center line and struck their vehicle. Several months later, two football players in Cache County were killed when the driver was trying to send a text message and drove into oncoming traffic." (In fact, Terryl says she was relying on a law enforcement tip that the driver was sending a text message, something that was never proven. Similarly, Terryl had heard widely circulated allegations that the driver who killed Lauren Mulkey was on the phone at the time of the wreck, but that was never proven in court, and the driver pleaded guilty to negligent homicide.)

Quite obviously, Terryl was trying to counter criticism that texting and driving was merely a theoretical problem, and to prove that it was fair to compare it to such problems as drinking and driving.

"According to our office," Terryl said, putting a fine point on it, "we have not had a DUI homicide since 2001. But in the past two and a half years, we've had four deaths due to text messaging."

Terryl listed a few more stats, including research from the state of Utah that the number-one cause of distraction-related wrecks in the state owed to cell phone use by motorists. She cited some of Dr. Strayer's research. The committee seemed interested enough, but this was, in a way, more of the same. Each side had their advocates.

Terryl noticed a few legislators were texting, including Wimmer.

Terryl concluded: "I have read we should concentrate on DUIs, that we shouldn't regulate what people do in their cars, and that business people should be able to conduct business while in their vehicles. Those thoughts are unacceptable knowing how dangerous a text-messaging driver is."

She sat.

UP STOOD A YOUNG woman, Paula Hernandez, a high school student, who spoke only for around a minute, saying that texting and driving could be dangerous and the law should "take that privilege away."

Committee members thanked her for her courage, ready to move on.

"Anyone else from the audience?" Aagard asked.

A YOUNG MAN STOOD up.

"Sir, please come forward."

The young man walked to the front. He had a contradictory physical presence, taking up space, a decent-sized kid, but hollowed out. Slumped from the inside. He wore a dark suit and a tie.

"Again, if you can state your name and who you're with."

"My name is Reggie Shaw and I am a citizen."

Up until this point, the most the quiet young man had said about the accident and his feelings was to the probation officer, Kaylene Yonk. He paused, and looked up.

"I'm from Tremonton, Utah." He got that far before his voice broke the first time. "And, uh, the lady that just spoke mentioned of, uh . . . an accident caused, um, in September 2006. Um, I was the one driving the car, texting while driving. Um. Excuse me, I apologize, this is hard for me," he said, his voice cracking, on the edge of grief. "At the time, before, I did not know of the dangers. I was young. I was ignorant. No one had really talked to me about it, and I know that's how a lot of people feel today. A lot of people might not know what text messaging is, what is involved, how you do it. It is dangerous."

By this point, Reggie had regained much of his composure. It seemed that when he talked about the theoretical, the policy—like whether texting is dangerous—he could hold it together. When he spoke about what he had done, how he felt, then he fell apart, as he did with the next sentence.

"That accident has changed my life forever," he said, now really fighting to hold it together. He choked back a sob. "Never, to this point, have I gotten a chance to apologize to those families. I know that they're here today, and I'd like them to know I'm sincerely sorry."

In the audience, Terryl and the families were not sure what to think. Who was this kid, exactly? The one who had lied? The one trying to do a number on Judge Willmore to get his sentence lightened? Was he just satisfying his community-service requirement as outlined in the plea in abeyance? Or was this something more sincere?

Terryl had met him briefly, prior to the hearing. She, in particular, was having a hell of a time sizing him up. An emotional wall that had been around him came down. She picked up on it when they had met, and now, here was Reggie, sounding a lot more like the person Kaylene had described.

"This accident has affected my life forever. Um, I can't even put it into words. And to see a law passed that would prevent people to do this would mean a lot to me, to be able to know that nobody else would have to go through what I've gone through. That they would be aware of the dangers that this text messaging is, and what it can do, and the effects it can have. So please listen to the things that I have said, and know it is dangerous and it affects a lot of people's lives and it is not safe."

He turned and walked back to his seat.

There was a hush over the chamber. Terryl looked around and noticed something: Reggie had brought down the house. "There wasn't a dry eye," she says. All the texting had stopped, all the fidgeting. "You could hear a pin drop."

TERRYL COULD FEEL HER perspective radically change. It wasn't just the words or the tone. Reggie was deeply injured, you could just feel it. "In a moment," she says, "I completely turned, in a moment."

It was a remarkable change for her, as instant as the speed with which she'd turned her energies against Reggie. She had heard this apology and it sounded real. She wasn't looking to punish Reggie for the sake of punishment, but she couldn't stand someone who victimized other people and just acted like it was the victim's fault, or there was no blame. That was what bad parents did to defenseless little girls. Now she heard Reggie sound completely real, tortured even. He sounded truly sorry.

"What he said resonated so much. He could've been anyone's brother, anyone's boyfriend, anybody's child—Taylor, or Jayme, any of us. When he just sobbed, I thought: 'You could be anybody.'"

She still wanted him to serve jail time and she wanted something good to come from the tragedy. But she realized that she and Reggie were no longer on opposite sides. Maybe they could join forces.

"I decided then and there that I would work with him."

REPRESENTATIVE CLARK THOUGHT: *REGGIE just turned the whole thing around.*

"We all just sat there dumbfounded, like: 'Oh, my gosh, that could've been our kid,'" Clark says, looking back.

Aagard quietly said: "We appreciate your courage to do this."

The chair asked for comment from the members of the committee. There was a brief discussion. Something was different, something even more sober. The makeup of the committee had changed from the last meeting, too. Given all the absences, there were more Democrats than Republicans by a margin of 4–3, with the Democrats perhaps more likely to push the bill out of committee.

Still, going into the hearing, Clark had serious doubts. "I'd been lobbying them all along, and I couldn't get any commitments from them at all. They said things like: 'Well, you do that and pretty soon you're going to take the phones away from us.'"

After Aagard opened the discussion, Wimmer asked to speak. He reiterated his concerns, but he changed the emphasis of his opposition to whether a law like this could be enforceable. He proposed to do a study. "Let's get a report back a year later, in the next session, on if this is effective," he said. "In my experience as a police officer, I really question whether this is enforceable at all. I'd like to be proven wrong— who hasn't been stuck behind a person texting and wanted to throw the phone out the window." The line got only a mild laugh. The room had changed; Reggie had altered the tenor to something decidedly sober.

Wimmer then said something remarkable. "I hope we're very careful as we go down this road. I'm adamantly, adamantly opposed to outlawing cell phone usage while we're driving. I hope we don't start carving out every little possible thing that I'm sure ninety percent of us have done in our vehicle—adjusting the radio, adjusting the air conditioner, all the careless distractions we do."

It was noteworthy because it was a sign of what had happened to the room, and what was about to happen. A few minutes later, there was a motion to pass the bill out of committee. It passed, with only Wimmer objecting. It was a huge hurdle.

On February 26, it passed the full House, by a vote of 55–20. It would still have to pass the Senate and be signed into law.

CHAPTER 41

JUSTICE

ON MARCH 10, AT 3:50 p.m., justice was served in a maelstrom of emotion, grief, and a nagging uncertainty that had plagued these people since it all began.

Judge Willmore sat elevated, sedate, tie barely showing over the top of his black robe, his torso seeming nearly to shrink behind the bench. He spoke slowly, deliberately, a slight, distant whistle in his voice accompanying some of his harder syllables. Short-cropped salt-and-pepper hair, glasses, he was a patient, almost invisible arbiter of justice with an occasional strained pitch in his voice to punctuate the frustrations this case had inspired in him.

In the copy of *Les Misérables* that he kept in the upper right drawer of his desk, he had underlined many passages. There was one, on page 74 of his edition, that comes under the heading "A Place for Arriving at Convictions" and discusses a chaotic courtroom scene in France in the early 1900s, where the book's protagonist, Jean Valjean, faces judgment.

"At one end of the hall at which he found himself, heedless justice in threadbare robes were biting their fingernails or closing their eyelids, at the other end was a ragged rabble. There were lawyers in all sorts of attitudes," the passage begins, and goes on to describe this mess of humanity, all the absorbed and self-absorbed participants, even the inattentive judges, but then concludes: "for men felt herein the presence of that great human thing which is called law and that great divine thing that is called justice."

Judge Willmore called to order the courtroom. Seated in the galley, a smattering of onlookers joined the families. On the left side (from the judge's perspective) Terryl was seated beside Jackie, who sat with Megan and her husband. Leila sat on the other side of the aisle.

Linton waited at the prosecutor's table, in a black suit and maroon tie. Beside him was Rindlisbacher in a heavy, brown highway patrol jacket. Reggie and Bunderson sat to the far left; Reggie looked less terrified than brave, almost welcoming.

Judge Willmore explained that a plea had been reached, and that he'd rule shortly on the particulars of the sentence. He had a few conditions up the sleeve of his robe that the parties hadn't bargained for.

First, he asked Bunderson and Reggie to come to the defense table, across the aisle from Linton, to hear what they had to say in advance of sentencing. Bunderson, half a head shorter than Reggie, pulled the thin microphone close and began by asking for Judge Willmore to set another hearing in six months. By that time, Bunderson said, Reggie should have been able to complete whatever jail time he'd been assigned and the fifty hours of community service that the parties had agreed to.

He's not done yet, Bunderson said, but "he's heading that direction." And in six months, Bunderson hoped, the judge would allow Reggie to be done with his sentence and have his record cleared.

Then the lawyer delved into the presentencing report, the one compiled by Kaylene Yonk. "I've rarely seen a presentencing report so favorable to the defendant." He went on to say that Kaylene described Reggie as having been in shock after the accident and that's why, Bunderson said, he didn't tell Rindlisbacher what had happened or even remember or know what had happened.

"He's not a criminal in any sense of the word other than he was involved in an accident and he's willing to take responsibility for it," Bunderson said. "He's willing to man up."

And he sought to distinguish what Reggie had done—the act of texting and driving—from other negligent behaviors with terrible consequences. "This is not a crime in the sense that you deal with every day or in the sense that Mr. Linton and I deal with every day. There's no intent in any of this.

"Your Honor," he continued, becoming more animated, "there but for the grace of God go I, go you, goes everyone in this courtroom—Mr. Linton, Officer Rindlisbacher, and everyone else here. We've all looked at something or done something to distract ourselves while driving."

He seemed to address the courtroom. "Any of you who have yelled at the kids, you've been distracted. Any of you who looked at a flock of geese while driving, you've been distracted. It's just a matter of degree."

"It's a matter of consequence," the judge interrupted, a slight edge in his voice.

"I'm not denying that Your Honor," Bunderson countered quickly. "This could have happened to any of us."

DIRECTLY BEHIND BUNDERSON AND Reggie, a few rows back, Jackie jotted notes on a small pad, looking impassive. Megan looked irritated, even bored, her eyes sometimes wandering from the proceedings. Their faces didn't betray their incredulity. Bunderson, in describing Reggie's actions as being like any other distraction, seemed to run afoul of everything that they had learned during this twenty-nine-month period, from Dr. Strayer and all the rest. To these families, it all seemed so disingenuous, the thinly veiled plea for mercy and, most of all, the idea that Reggie was taking responsibility, "willing to man up."

All they'd had for two and a half years was silence and lies. Now, a veritable deathbed conversion, a presentencing conversion. How contrite or sorry could Reggie really be, and why?

Bunderson concluded. Judge Willmore asked if Reggie had anything to say.

REGGIE HAD STARTED WRITING apology letters days after the accident, the ones he never sent. Some just said "I'm sorry, I'm sorry, I'm sorry," over and over. He'd practiced this idea in his head virtually daily, multiple times a day. But when he started, the first few minutes, it sounded like a high school essay—topic sentences followed by explanations.

He began: "I wrote this letter to those I have harmed."

He wore a blue suit and yellow tie, his hair close-cut, the sideburns slightly long as had become the style. He'd filled out the last couple of years, looking just shy of imposing, especially next to the older, slighter Bunderson. His voice tended to be slightly high, plaintive.

"At this time, I would like to express my love and remorse toward their families who have been in my heart and in my prayers daily," he read from his letter. "I truly realize and I understand that I was wrong to not pay closer attention."

Behind him, Jackie shifted in her seat but seemed unmoved.

Reggie apologized for his ignorance, saying he wished he could rewind the tape, give back the lives of "these great men." He talked about learning of the dangers of distracted driving, vowing to "take this terrible situation and help make the roads a safer place for us and our families."

Here, a few minutes in, the tenor changed, almost like he'd completed the part he felt compelled to say, or was coached to say, and now moved on to the part he wanted to say.

"I've learned over the past couple of years . . ." Reggie paused to not lose his emotions. ". . . that good people strive to care for and help protect people's lives. Good people learn from their mistakes and try to make them right and seek forgiveness from those they have offended or harmed.

"I am here today striving to be a good person.

"In reflecting at who I was age at nineteen, I am very ashamed by my decisions and my decision making at that time." A bailiff walked over and put a box of tissue in front of Reggie. Behind him, in the rows, Megan crossed her arms, Jackie's head was tilted as if watching an unusual animal at the zoo.

"I am also very embarrassed by how I handled myself in this situation. I give you my solemn vow that I will never behave in that manner again." Reggie choked back tears. Bunderson looked straight down. Reggie explained he regularly cries and that his "tears of remorse have caused me to change in every possible way."

He continued: "I can't take back my actions. I can only live from this day on in service to others in remorseful remembrance of James Furfaro and Keith O'Dell." He could barely get the words out, the next few minutes halting in presentation, and in that way, and substance, the high-school-essay nature melted away to core grief.

He explained that remorse, to be real, must be "coupled with action." The time has come, he said, to act. First, he said, he must apologize.

Hardly able to speak now, he looked up, briefly. "I am so sorry. I am sorry for being careless and for not paying more attention behind the wheel that morning that led to the deaths of your loving husbands, sons, brothers, uncles, and, most importantly, your fathers. My apologies are sincere and my love for you is great.

"I beg and plead to these families for their forgiveness. I made a mistake, and although I realize that nothing can replace the lives of these great men and the pain these families have felt, I'm willing to do anything to gain their forgiveness."

It was full-body contrition. Behind Reggie, the families seemed unmoved, except Leila. She cried, but not because the young man apologizing had reached her but rather because his words, hollow to her, could not bring Keith back.

Bunderson continued to look down, boring a hole in the table beneath him.

"I have felt pain and heartache in unexplainable great measures the last few years. But I know my greatest pain cannot compare to the pain they felt after losing a loved one," Reggie said. "I am willing to do anything to make this right to you. I'm so sorry. I truly am." He sobbed, barely able to speak. He bit his bottom lip, a desperate measure to hold his body together at the mouth. "I can only hope and pray that I can continue to live my life to make things better forever for everyone. I shall forever live my life connected to these families by the weight of my foolishness. I am truly sorry.

"I can finally be the person I should've been as a nineteen-year-old kid," Reggie said.

He looked up.

"From now on, my purpose in life is to make others aware of dangers that ultimately took lives on the morning of September 22, 2006."

FILLING THE SILENCE THAT followed Reggie's statement, Bunderson offered a quick comment, a reiteration: He asked for a follow-up court date in six months, so that the judge could sign off that Reggie had finished his community service, jail time, and that the young man could move on.

It felt mildly off-key after Reggie had pledged his life to service. But, then, there are courtroom promises and deathbed conversions, and then there's reality.

THE JUDGE ASKED LINTON whether any victims or family members wished to speak. The first to rise was John Kaiserman, wearing a black vest over a red shirt and sporting his customary handlebar mustache. He offered apologies and condolences to the families of Jim and Keith. He then talked of his own journey since the accident, which he says left him with injuries that forced him to give up his passion of being a farrier. "How great the day when you can enjoy the labor that puts the food on your table," said Kaiserman, with a Shakespearean flair, a sprinkle of melodrama, jutting in and illuminating every few paragraphs of his otherwise plainspoken delivery. He wiped away a tear and spoke of how his wife and daughters, since the accident, see him walk out the door each day and say: "Be safe, Daddy."

He questioned Reggie's sincerity. "The correct time for dialogue was September 2006. A great deal of pain and countless hours of wasted time and energy could've been avoided. At this time, the apologies you make seem to be heartfelt but they ring fairly hollow."

And he told Reggie how this incident has divided a community, with some people thinking the sentence is too harsh, that this was a mere accident, while others think it too light and that "nothing short of your head on a platter will suffice."

"Reggie, I hope in the future you may be able to get past this. Somehow everyone who survives this will."

JACKIE SPOKE NEXT, SHAKING while she read quickly as if in great discomfort, seemingly more nervous than grieving. She chose the moment not to admonish Reggie, but to celebrate Jim, talking of her husband's zest for life, the unicycling, geocaching, Dance Dance Revolution, World of Warcraft, how his girls miss him and she misses him with the girls. She talked of his many accolades as a rocket scientist. "I'd planned to grow old with Jim. We were supposed to raise children together, supposed to enjoy life together. Today, I'm alone, a single parent trying to raise the kids by myself. All we have are faded memories."

She concluded by telling Judge Willmore: "Please send a message to people everywhere that texting and driving are dangerous and the effects can be both heartbreaking and devastating."

"MY DAD WAS THE most important part of my life," said Megan, who spoke next. "He always helped me.

"My wedding was the hardest thing ever, having to walk down the aisle without my dad."

She mentioned all the times that they spent together.

"I don't know what else to say."

LEILA, HER BLOND HAIR long and straight to the middle of her back, came to the podium on the verge of tears. But there was force and dignity in her even as she, acknowledging her public grieving, said, "Forgive me. I've been very emotional for twenty-nine and a half months. Forgive me."

Like Jackie did of Jim, Leila celebrated Keith, "the sweetest, kindest, quiet, shy genius." She talked of his hobbies and talents, how he was celebrated at his funeral as one of the great rocket scientists. She spoke of his closeness with Megan, and, in conclusion, of his relationship with her, and hers to him.

"He was my best friend for over thirty years and I will miss him forever."

LINTON ADDRESSED THE JUDGE, saying this had been one of the toughest cases, "perhaps more difficult than cases gotten more notoriety and more attention—murder cases, rape cases. In all of those cases, I knew that somebody had intended ill upon another."

In this case, he said, the act was one that, as Bunderson said, "seems so commonplace."

But then he took a strong stance against something the opposing counsel had said.

"I owe it to this community to say that texting while you're driving is not the same as looking at geese on the horizon. It's not the same as adjusting the radio. It's a singular act of criminal negligence. That doesn't make Mr. Shaw evil—I would never say that—but it leads to a tragic result, and it is in an effort to avoid this kind of tragedy again that I brought this charge, Your Honor.

"This is criminal conduct not committed by evil people but committed by people who forgot they're traveling down the road with two thousand pounds of steel beneath them with a potential consequence of absolute and total devastation."

IN A VOICE JUST shy of quiet, Judge Willmore asked Reggie to come to the podium.

"What a tragedy," the judge began. "It's a struggle for me to know what to do in a case like this."

He said his heart goes out to the families and thanked them for their letters and statements.

"To you, Mr. Shaw," he said, turning to Reggie. "What you did was a crime. It wasn't a mistake. I keep hearing 'mistake, mistake, mistake.'"

With his quiet, penetrating intensity, he said that it's not a "mistake" if a reasonable person knows the risks. "I can't see it any other way!" The words burst from him.

Then, just as instantly, his calm returned as he spoke of his frustration that the case wore on for so long. "It should've never taken this long," he said, and gave an oblique slap to the lawyers, noting "our system is attorney-driven," without elaborating.

He talked of being vexed by Reggie's lack of apology, saying he could've expressed sorrow to the families without admitting guilt. Absent an apology, he said, "common decency is thrown to the wayside."

He turned to the actual sentencing. As a preface, the judge explained that the law allowed a penalty of up to a year in jail for each count of negligent homicide. However, he noted, the deal reached by the attorneys restricted him to sentencing Reggie to between fifteen and ninety days, a restriction he said frustrated him and his ability to exercise the full of his judgment.

He then thanked Kaylene for what he characterized as her excellent probation report, reminding the courtroom it recommended a sentence of fifteen days. He contrasted her view, and that of some others in the case, who wanted a more lenient sentence, with the view of Rindlisbacher, who would like to see the full weight of the law brought to bear.

"I've never seen such a conflict in law enforcement," he said.

And he added: "It just shows how society views this crime."

It's a pointed example of Judge Willmore's understated eloquence, an easy thing to miss, but a critical insight. This case, he was saying, happened on a razor's edge. Some wanting charges, others not, but all of them, in one way or another, seeing themselves in Reggie, viewing this tiny moment in time as a projection of how they would've handled themselves, or have. His attention, ours, is so fragile. It can happen to anyone. Can any of us be expected to know the consequences of actions that feel so close to mere accident?

The judge said he feared that some will view the decision he is poised to announce without understanding the severity and complexity of distraction science. "The public will judge this sentence based only on the media—very few of the public will do anything to educate themselves about this problem."

Earlier in the day, the judge said, he called the Utah State Legislature to ask the status of the bill Reggie had testified about. He was told it is in a Senate committee and likely to pass. "A big part of that is due to the testimony of Reggie Shaw."

At the podium, Reggie wiped away a tear.

"Mr. Shaw, that doesn't make you a hero," Judge Willmore addressed him. Reggie shook his head. "All it shows is that you're trying to make something good out of this tragedy.

"I was going to sentence you to ninety days," the judge said. In light of the legislative action and Reggie's other efforts, the judge ordered thirty days of jail time, which could expand to ninety days if Reggie didn't meet his other conditions:

- Two hundred hours of community service, including, specifically, 150 hours speaking to local schools about distracted driving;
- Continuing to work with legislators.

He added a final, unusual piece to the sentence.

"The last condition, and I'll get beat up in the media for this, is that you will read *Les Misérables* by Victor Hugo, which talks about a man who has done a terrible wrong and makes it right again."

The story starts when Jean Valjean is released from prison after serving nineteen years for stealing a loaf of bread to feed his starving family. Just out of prison, he steals again, this time silver candlesticks from a shelter. He is caught by a benevolent bishop, who keeps Valjean from going to prison for life by telling police he'd given the squalid ex-prisoner the candlesticks as a gift.

The bishop urges Valjean to take the candlesticks to invest as seedlings in a life devoted to good. He does just that, becoming a wealthy benefactor of the impoverished, a generous and kind man beyond compare. And yet he remains haunted and hunted, his past troubles following him like a mark of Cain.

With this parting decree, Judge Willmore dismissed the court.

Leila and her family wandered out. Jackie lingered. She numbly accepted apologies from the Shaw family.

Linton saw Reggie walking in his direction, maybe on purpose, maybe just blank-faced, heading in that direction. Linton couldn't tell.

"Reggie."

The shell-shocked young man looked up.

"This is your true mission. Your true mission may be saving lives."

Reggie's face was blank.

"Instead of trying to convert people, your true mission is saving lives."

Linton recalled Bunderson rushing over, apparently angry that the prosecutor would be talking to his client. Reggie kept looking at Linton.

"Either he agreed with me," Linton says he thought, "or it was just so true you couldn't get around it."

CHAPTER 42

THE LAWMAKERS

TWO DAYS LATER, ON March 12, the Utah State Senate passed legislation to ban texting and driving. The vote was 26–1, not a surprise at that point, given that the Senate, while still conservative by any national measure, was less so than the House, and certainly less than the law enforcement committee in the House that the legislation passed through a few weeks earlier.

The legislation, which had been shepherded by Senator Hillyard, made it a class C misdemeanor to use a device to text or send email, and a class B misdemeanor if a driver doing such an act were to cause severe bodily injury. It would be a second-degree felony if death resulted.

The standard was "simple negligence," as the law puts it: "a failure to exercise a degree of care that a reasonable and prudent individual would exercise under the same or similar circumstances."

On March 25, Governor Jon Huntsman signed into law the toughest ban in the country on texting while driving.

Utah residents, if they hadn't been already, were put on notice. This is dangerous and illegal, and they had better know it.

People who followed the politics and policies of the state were somewhat astounded. They also knew what deserved the credit: the deaths of Jim and Keith. Scott Wyatt, the former county attorney in Cache County, the man who hired Terryl, says she "deserves most of the credit" for the law. She pushed the prosecutors to go after Reggie, then took on the cause of texting and driving, amassing a binder of evidence for media and legislators.

Dr. Strayer knew well the long odds something like this would pass through Utah's house. He joined others in saying that equal, if not more, credit also goes to another person who had begun to turn tragedy into redemption.

Says Dr. Strayer: "They should call it 'Reggie's Law.'"

CHAPTER 43

JUSTICE

A FILM CREW VIDEOTAPED Reggie's entrance into jail on May 8. The crew worked for Zero Fatalities, a public safety organization aimed at curbing automobile accidents and car-related injuries and deaths.

In the video, Reggie's hands are up against a wall as a guard pats him down. He wears jeans and a blue sweatshirt. Moments later, the video cuts to him exiting the "Inmate Changing Room."

Now he wears a white-and-orange-striped shirt and pants. He looks stocky, the years since the accident having turned him from adolescence into a man. A voice-over accompanies the images. It belongs to Terryl: "To Reggie, who has never been away from home unless it was for something positive, every day will seem like an eternity. He will be with the general population so he will be with people who have long criminal histories."

As Reggie hands over his old clothes, Linton's voice comes in: "He will be in with some very unsavory people."

THE FIRST NIGHT, THERE was no room for Reggie to have a bed. So he received a mattress put on the floor next to a bunk. He didn't sleep.

He didn't sleep for days. He tried to keep his eyes down. He shared a cell at some point with someone who had nearly beaten his girlfriend to death. How scared should Reggie be? How self-pitying? He'd actually killed someone.

THE CASE STUCK WITH Judge Willmore. He'd stopped using his phone while driving. Midway through Reggie's sentence, he decided it was enough. In a highly unusual move—again, reflecting a highly unusual case—he called Linton and Bunderson and asked if he might shorten the sentence; in effect, commute it, with the understanding that Reggie would serve all of it, and more, if he didn't meet his obligations.

Reggie served eighteen days of his thirty-day sentence.

CHAPTER 44

REGGIE

IN LATE SEPTEMBER 2009, almost three years to the day since the accident, Reggie boarded an airplane for Washington, D.C. He was going to attend the first ever Distracted Driving Summit, which was organized by Secretary of Transportation Ray LaHood.

Secretary LaHood, tall and genial, a former Illinois congressman, had begun focusing himself, and his agency, on the issue of preventable traffic fatalities, especially those caused by use of technology by drivers. "This," he said, "is probably the most important meeting in the history of the Department of Transportation."

Held at a downtown hotel, the conference drew scientists, public safety advocates, and legislators, and it brought together people involved in distracted-driving accidents, both victims' families and, in one case, the driver in a deadly wreck. That was Reggie. The event included all the big ideas and some of the same players as Reggie's trial, but played out on a much bigger stage.

Reggie told his story to the attendees. "Two men, who were fathers, and also husbands, cared for their family, wanted the best for them. And because of my choice to text and drive, I took their lives. I changed the lives of these families. I changed my life forever," he said.

It made a strong impression on Secretary LaHood.

"His story was so compelling, the way he tells it, the way he gets their attention," the secretary reflects. He was unlike most of the other people who testified about such dangers; they tended to be researchers, victims, or their families. "It's different when you hear it from someone who has committed the terrible act of killing someone."

I WAS AMONG THOSE people at the conference, though by that time I was no stranger to Reggie's story, and his passion. As part of a series about distracted driving, I'd written a front-page story about Reggie that appeared just a few weeks before the summit in Washington. Even in getting to know Reggie and Terryl in the past month or so, I could tell they were deep wells—open, accessible, injured people.

It was also clear that the topic of distracted driving went well beyond the accessibility of technology in cars. It was a story about science, primitive instinct, and culture and sociology, and the tension between the need to stay connected and the risks of being entranced by their siren call.

During the conference, the *New York Times* published another long story in our "Driven to Distraction" series about people who feel compelled to get work done while they're driving—"mobile workers." Under the headline "At 60 M.P.H., Office Work Is High Risk," the story quoted businesspeople describing the desperate need to respond to client inquiries within seconds, or risk losing a sale.

The story contrasted those demands with the science showing that when someone was being responsive, it wasn't the same thing as being effective. In fact, not only did the research show that it is physically impossible to do more than one thing at a time, but the mere effort to juggle forces a person to switch tasks in a way that robs performance from each task. Another study, done in 2006 at the University of California at Los Angeles, compared brains of people trying to multitask to those doing a single job and found something fascinating: People who focused on doing only one thing performed learning and memory tasks with the hippocampus, the brain's memory center, whereas people trying to do two tasks relied more heavily to learn and memorize on the parts of the brain associated with motor skills. That, the researchers said, was not a good way to enforce long-term learning and memory.

When someone is doing two things at once, "there is an illusion of productivity," David E. Meyer, a professor of psychology at the University of Michigan told me for the story. His research had showed that performance drops sharply when someone multitasks. "It's actually counterproductive.

"To the extent that someone is focused on driving, the quality of work product is diminished," he added. "To the extent someone is focused on work and not driving, there's a risk of crashing and burning. Something's got to give."

Sometimes, the results were tragic.

The story described the Coca-Cola truck driver who, distracted by the computer he used to get his delivery orders, killed a seven-year-old in the backseat of an oncoming truck. In the instant after the wreck, the boy's mother went chasing after the truck driver. We quoted her as recalling:

> "I said, 'Why, why, why?'" she recalled screaming at him. "He told me, 'I just took my eyes off the road for a second because I was looking at my computer.'"
> She started chasing him.
> "I went into a mad rage," she said. "If he'd said he'd fallen asleep, maybe I'd have understood. But using a computer?"

A few days later, we wrote a piece about truckers in general who insisted they needed to text while driving to stay in touch with dispatch. The story described the concerns by the trucking industry that if texting were made illegal, it could hinder their efficiency of communications.

AT THE END OF the distracted driving conference, Secretary LaHood announced that President Obama had signed an order forbidding federal employees from texting while driving during work hours. The secretary declared that other measures were being considered as well. He added that the order to restrict text messaging by federal employees behind the wheel "sends a very clear signal to the American public that distracted driving is dangerous and unacceptable."

IT COULD HAVE BEEN the end of it for Reggie. He'd helped pass a law, spoken at numerous events, put himself out in front of three hundred people in Washington, D.C., at the conference organized by Secretary LaHood.

At the end of September, Bunderson wrote a letter to Reggie and his parents. It began:

> *Dear Reggie, Ed and Mary Jane:*
> *Just for your files, enclosed you will find a copy of the Order of*
> *Dismissal.*

The letter went on to discuss a few finalities involving insurance and Mr. Kaiserman and to explain that Reggie's file could be expunged and the court records sealed. It would cost between $500 to $850.

He enclosed his latest billing statement.

He concluded:

> *Reggie, I wish you good luck in all your future endeavors, and*
> *feel free to call me if you have any questions.*

It was really, and finally over.
But it wasn't. Not even close.

THAT SAME MONTH, REGGIE had been writing his essay about *Les Misérables*, explaining the significance to him of the epic story. He submitted it to Judge Willmore in September.

The three-page, single-spaced report reflects a great deal of analysis. The use of language is anything but inarticulate, as Bunderson once feared Reggie might be perceived if he testified. It is flowing and sophisticated, beginning with a brief passage summarizing the plot. Jean Valjean, the book's hero, is shown mercy by a bishop for a criminal act. The bishop, Reggie wrote "taught him a lesson without directly saying anything to him.

"Valjean did not promise anything to the bishop at this time and the bishop didn't expect him to. Valjean did not promise to him to spend his life in service, but it was after he left that he looked inside of himself and decided that he wanted to help others and this life would be better served in helping others.

"I think this is the message that I'm supposed to receive."

He returned to summarizing the book, emphasizing examples of how Valjean served others. Then Reggie applied the story to his own life. "I now understand that for me to live in any matter other than serving others would be selfish and unjust. I don't need to go around making a promise that I'm going to devote the rest of my life to service," he wrote. "But through this terrible situation that I have created, I have received a gift that I have the ability to affect a lot of people's lives. I can make a promise to myself that I will do whatever it takes to try and help people by making sure that nobody makes the same mistake that I did.

"Valjean has taught me that any person in any situation has the choice to change. Change is a choice. It's that simple. Much like he changed in the story, I have changed and will change as I commit myself to service toward society."

He finished by describing the lesson he would "take most from the book." Namely: not seeking profit or praise from his redemptive efforts. "When I go out and hopefully have an impact on society, I will always remember that I don't need praise, in fact, not only do I not need it, but I don't want it."

IN JANUARY 2010, REGGIE appeared on an *Oprah* segment about driver distraction, telling his story.

The night before the taping of the show, in Chicago, Reggie had dinner with Terryl and Megan O'Dell. Reggie and Megan made a decent connection, what Reggie hoped would be the beginning of a friendship. The two started talking on a regular basis.

The requests poured in for Reggie to talk publicly, and even though his sentence was complete, he said yes, over and over. And each time, he displayed the same level of grief, what people who heard him would characterize as inescapable sincerity.

That June, he appeared in front of the most elite basketball players in the world, an audience of one hundred rookies entering the National Basketball Association. The NBA had heard about Reggie and wanted him to reach out to its players because the league felt that so much pressure was mounting for them to be connected all the time. They were young, with phones and fast cars, and, adding to all of it, the emergence of social media. "Our players are very active on the social media front, they like to keep their fans abreast of their progress, something as simple as the fact they're going from the breakfast table to the workout room, and sometimes larger issues," says Rory Sparrow, a former NBA player who had become vice president of player development.

A roomful of NBA stars, young and on top of the world, might have intimidated anyone, let alone Reggie, a good basketball player himself, who aspired to coach and lead players. But far from daunted, he charged right in. He gave a version of his usual presentation. He asked them: Would you pause to text while driving to the basket? And then he cried as he described his guilt, and pleaded with them not to "become like me."

The roomful of players was silent, rapt; then, when Reggie was finished, they stood in applause.

Sparrow says: "As his story comes out, you realize that, every time, he is reliving it."

CHAPTER 45

REDEMPTION

M Y NAME IS REGGIE Shaw."

The cavernous auditorium seems like it could swallow Reggie, who stands alone on stage. He holds a microphone in his right hand. He wears a tie.

"I'm going to tell you a story."

Around five hundred students from Box Elder High School watch and, to a greater or lesser extent, listen. Sprinkled throughout, students play games on their phones. It's a midmorning assembly focusing on safe driving.

"Keep in mind that, me and you, we're not different," Reggie says. "I've been to Brigham City millions of times. I've played basketball in your gym. I've played football on your field.

"When I was young, growing up, I always thought I was invincible. Growing up in . . . Tremonton, nothing bad happened to me, or people in my life." Reggie shifts from foot to foot. "I turned nineteen, got a job in Logan, painting houses over there. I got up early and headed over the mountains there.

"One morning, I got up, headed to work, and I made a choice, made a decision that day, something I did all the time. I decided I was going to text and drive. On my way to work, I was reading and sending text messages. I go across the center line, and I hit another car."

For the first time, Reggie pauses. He licks his lips, swallows. He sports a goatee, no longer the person with the bowl haircut and boyish face that shone on his driver's license. It's a Thursday in early April 2013.

"On the way to work were two men with families they loved and cared for, they provided for. They were both killed on impact. . . . I live every single day of my life knowing I took two men's lives in an accident that was one hundred percent preventable."

Reggie's parents sit in the front row. Ed wears a red short-sleeve shirt, work boots and jeans, eyeglasses; his hair is starting to gray. Mary Jane sniffles, her legs crossed; together, Reggie's parents seem barely to be holding it together, folding in on themselves. Reggie's still relatively in control, reciting facts and emotion. But Ed and Mary Jane know what's coming. Reggie's going to crash, and he's going to take these half-attentive students with him.

"On my sentencing day, the last day of court, the families got a chance to speak to me, to address me. One of the men has a young daughter, about my age, and she gets up and talks about how she'd just gotten married and she talks about her wedding and she talks about how her father was not there to walk her down the aisle." Reggie sniffles. "She turns to me, and she looks me right in the eye, and she says: 'Because you took that away from me.' She was absolutely right. I wanted to text and drive, and I took that opportunity away from her, to have that with her father. I was thinking, at the time of the accident, I was only thinking about that message. I want you to think about every single message you've ever sent or received. Is there any one message you'd take your life or someone else's life for? Absolutely not. It's been almost eight years since that accident happened. It is difficult for me every single day. Not a day goes by that I don't think about that accident and say to myself: 'I wish I could go back and change what happened.' I go to bed every night, and look at myself in the mirror every morning, knowing I caused an accident that was one hundred percent preventable. I took two great men's lives."

Reggie's words aren't ornate or filled with rhetorical flourish. Their power, building and building, comes from grief. It is changing the physical surroundings, now threatening to swallow the auditorium. Reggie's choking it back, straining to hold it. Students lean forward, a few blink tears.

"I only spent thirty days in jail. People ask: 'Do you feel like that's enough time—for what you did?' I don't know. Still to this day, I think about that question. I don't know." Reggie stands up a bit straighter, takes a deep breath. This part he can handle. He tells them about the first three nights and four full days in twenty-four-hour lockdown. He tells them how he slept on a "boat," which is a small blue bed, an inch and a half thick, because the jail was overcrowded and the bunks were taken by his cellmates. "They throw it on the floor and hand me a thin blanket, shove me in, and shut the door and say good luck."

For the first four days, he explains, he sat on the cement floor twenty-three hours a day. Then he got his bunk. His cellmate was being held awaiting transfer to the state penitentiary for beating his girlfriend nearly to death. Reggie's holding the microphone with both hands.

"The thing about jail is nobody cares about you. Nobody liked me there. They called me the 'Textual Offender.' I go back to that question," he says, referring to whether he spent enough time in jail for his crime. "I don't know. One thing I can tell you: If I could go back and spend the rest of my life in jail to save those two men's lives, I would do it, in a heartbeat. I would live there every single day of my life rather than take two men's lives."

Now he's crying.

"I'm here for one reason. That's for you guys to look at me—" Reggie's choking back tears. He can't talk. He fights to finish the sentence. "And say: 'I don't want to be that guy.'"

There's just Reggie now, and his grief, five hundred students absorbed in it. He continues about how he doesn't want them to be him.

"'I don't want to put people through what he's put people through. I don't want to go through what he went through.' I can promise. I can promise you . . ." He starts to regain himself. "When you get behind the wheel of a car, that is not the time to make phone calls, to send text messages. Your friends can wait, your family can wait.

"Driving in the car, you thought a text was more important than their son or daughter. It is not worth it. It is not worth it. Please make a pledge with me today. Help me out. Please put your cell phones away when driving. Turn them off and put them away. You're going to save somebody's life."

Reggie takes a deep breath, a wind-down breath. Then, he just sums up, "I'd love to answer your questions. I'm very open about my story. Nothing crosses the line with me."

Applause rings out, and then it's done. Students file out, but a handful come to the front to thank Reggie.

"That was the bravest thing I've ever seen," says Nate Christensen, seventeen, a senior, wearing black shorts and a red T-shirt and a diamond stud in his ear. "I've done it," he says of texting. "I really didn't think it was a big issue. I looked up every few seconds and figured I'd be all right. That's obviously not the case."

RORY SPARROW, THE NBA player's rep, had characterized Reggie as "reliving" the experience every time he tells it. The NBA invites him back each year. He's spoken to the Detroit Lions football team, and gives a regular presentation to the top high school basketball players in the world when they gather at national clinics.

Sparrow says: "I'm not going to be so naive and say no one in the group ever texts and drives, but Reggie definitely changed attitudes."

Reggie has engendered respect, even from those who once fought him.

Linton: "I have never seen anybody try to redeem themselves as much as Reggie Shaw. Period. End of story."

Judge Willmore: "He's done more to effect change than anyone I've ever seen."

By 2013, Reggie had spoken at some forty different events for Zero Fatalities, the public safety group in Utah. Brent Wilhite, a program manager with the group, says it relies on people like Reggie who have been involved in accidents to put a human face on issues. But Reggie, in his own way, stands apart. "We've worked with people convicted of dangerous driving behaviors and they've been great ambassadors, but no one else has had the long-term commitment and passion of Reggie. He's gone so far above and beyond."

Wilhite says Reggie is good with an audience because he seems like an ordinary guy, the person who could be your friend or son, your boyfriend.

When it comes to finding a messenger, "we can't find anybody like Reggie."

Secretary LaHood, who in his four-and-a-half-year tenure as the nation's chief transportation authority, came across many harrowing tales and difficult safety issues. He says Reggie has really struck him with his commitment and testimony. The secretary invited Reggie to speak at his second federal summit, and then heard him at regional summits in Florida, Texas, and Illinois.

"He's a hero," Secretary LaHood says. "He bares his soul, he admits his guilt, and he tries to use that to persuade others to do right.

"Reggie has had a tremendous impact on people's thinking about distracted driving."

One could make a strong case that no single person has made as much of a difference when it comes to sending the message about the risks of texting and driving. Policy makers, like Secretary LaHood, who made driver distraction a national priority, certainly have had a huge impact. So have the early researchers, like Dr. Strayer, who were often sailing against headwinds generated by the corporations and profiteers benefiting from connected drivers. Among those three hundred people gathered at the distracted-driving summit, there were vigorous voices of people who lost family members, who created MADD-like groups to take on distraction.

But Reggie stands alone for changing a state law, taking his message nationally, showing relentless courage, putting himself, personally, on the line, to try to connect with the vulnerable age group of teen and young drivers.

Stephen Clark, the legislator who brought the bill to ban texting to the Utah House, eventually was honored by being named an LDS mission president in Missouri. His life now revolves around counseling young people and helping them digest their issues. He thinks Reggie has fulfilled a mission as great as any he could've aspired to through church work.

"If I was Reggie's mission president, and I was sitting across the table from him, I'd tell him: Your mission started the day you made that bad choice and caused that horrific accident. You started on your mission and you battled through the many, many challenges, the ups and downs, and you have been tried and tested and you have done all you can to make it right, and you have blessed the lives of many others who could be in the same position you are. Because of your willingness to stand forward and say 'I made a mistake,' you have fulfilled your mission in life. It was to tell the world that texting and driving is deadly. That's the missionary you have been."

He adds: "I believe Reggie has suffered enough. He needs to be able to forget it and move on."

I relayed this to Reggie as part of a long conversation we had on how he might be able to move on. When he heard what Clark said, Reggie started to cry. "Do I deserve it? Do I deserve to feel okay? What if I stop talking and making presentations and I feel better, or don't feel better, and then one day I wake up and read a story about someone who died texting and driving and I know I could've done something to stop it?"

That seemed to be the hardest part. Could Reggie move on? Would he allow himself to? Should he?

AND THERE WAS ANOTHER question. Reggie's many fans described him as a perfect spokesman because he is just a regular guy. Is he? Could any of us have done what Reggie did? Or was he more or less predisposed than the rest of us to lose his focus, take his eyes from the road, fiddle with his phone?

The increasingly sophisticated tools of neuroscience have something pointed to say about that.

CHAPTER 46

REGGIE'S BRAIN

A S I WAS GETTING to know Dr. Gazzaley in the spring of 2013, we sat down for lunch in the cafeteria at the UCSF campus where he has his offices. I told him about the accident and Reggie's story. Dr. Gazzaley listened quietly as he dug into a chopped salad.

I said I had a question for him: "Can I show you Reggie's brain?"

At first, he seemed a little taken aback.

"Sure," he said. Dr. Gazzaley's up for anything. Then he added: "He's just a regular guy, isn't he?"

Precisely what I was trying to find out.

BACK IN UTAH, DR. Strayer and a colleague named Jason Watson had been using advanced technologies like the MRI to look at attention from inside the brain. What networks are involved? What does distraction look like? While others in the field had been doing this as well, their work was more unusual in that it was part of a larger body of research aimed at exploring whether there are people who are particularly good at managing the onslaught of information. In other words, who can "multi-task" better than other people? Are there people who are worse?

Where did Reggie fit into the spectrum?

If, after all, he was an outlier, someone with a predisposition to distraction, it would be worth knowing. For instance, was there some signature on his brain of the concussions he sustained playing high school sports? Was there something odd in the way he processed information? Was he more likely to text and drive than the rest of us, or, when texting and driving, to be less able somehow to juggle the tasks?

In early April 2013, I drove with Reggie from downtown Salt Lake City to the University of Utah's neuroimaging center. The drive took us up I-15, the same route he'd driven with his parents when he came home from his first mission. Then we took the 600 South exit, heading east, in moderate, midafternoon traffic. Reggie said he was nervous, the idea of getting into the tube, having his head examined. Would they find something wrong?

On the other hand, this was the Reggie I'd come to know. He would turn down virtually no request that helped illuminate texting and driving—or its risks.

We almost didn't get to find out about Reggie's brain, at least not on that day. We almost wrecked instead.

Reggie was piloting his 2007 gold Mazda in the far right lane. In the lane to our immediate left was a sedan. In front of us, there was a pickup truck carrying stacks of folding chairs, the kind you put outside for a backyard barbecue on a sunny day, like this one. The chairs came loose from their bundle. We could not have been more than one hundred yards away, when the chairs began falling into our lane, setting up imminent impact. I looked quickly to the right, to the shoulder, but one of the chairs had fallen in that direction. To the left, the sedan had us boxed in.

I felt instant terror. I flashed on my kids and the idea that we would hit the chairs and start spinning around the highway, and I braced for impact. Reggie swerved left, narrowly missing two chairs rolling our way and sending us toward the sedan. Reggie veered smoothly back to the right, missing by a few feet the sedan, which was now just half a car length ahead of us. It was that close. Had Reggie not had his complete focus on the road, had he been lost for a millisecond, we'd certainly have crashed.

A few minutes later, dosed with perspective, we arrived at the neuroimaging center in a nondescript industrial park. Dr. Watson and Dr. Strayer were there to greet us. They walked us through the procedure. They'd take two kinds of pictures of Reggie's brain. One would be MRI images that would show us the structure of his brain. They'd also take fMRI, functional MRI images—so-called real-time MRI—which would measure blood flow changes as Reggie engaged in different behaviors. Specifically, they were going to watch what happened as Reggie tried to juggle tasks.

But before they did that, they wanted to know whether Reggie was the sort of person who was good at focusing and attending to information. So prior to putting him in the white tube, they asked him to sit in a small, windowless office, and take a written test. It involved solving some relatively complex equations, and also trying to remember information he'd been asked to memorize a few equations prior. This let them measure not just focus, but his short-term, or working, memory. What was Reggie's "baseline" ability to attend to information?

Then Reggie walked down the hallway, got into a medical gown, and climbed onto the MRI bed. The radiology technician made sure everything was in place. She put on his head the helmet that would allow him to see video images. She slid him into the tube.

THE FIRST RESULTS TO come back, the easy part, were the structural images. With Reggie's permission, Dr. Watson had sent them to Dr. Gazzaley, and he walked me through the images on his computer. It was a midsummer day, fogless. Through Dr. Gazzaley's window, across the quad, was the new Benioff Children's Hospital, a state-of-the-art facility funded by the largesse of Marc Benioff, who, along with his wife, gave $100 million to build the hospital from a fortune he made from his company Salesforce.com, which distributes and manages software for companies on the Internet. The convergence of technology, science, and medicine was on display everywhere you looked.

Dr. Gazzaley sat in a chair and called up a grainy black-and-white image. Reggie's brain. Just like he'd once before pulled up Mickey Hart's.

"Before we get to it, there's one thing I want to point out," he said. He drew my attention to the front of Reggie's skull. There was a little bump, a slight protrusion just in front of the sinus cavity.

"He's got a heavy brow."

There's nothing problematic about that—just an anatomical variability—though it might explain why Reggie sometimes looks as though he is leaning forward. Beside the image, Dr. Gazzaley pulled up the brain of another twenty-six-year-old—a random image taken from the lab—and there was no such protrusion.

Side by side, the two brains looked fairly similar. Dr. Gazzaley noted some very understated differences. Reggie appeared to have a slightly bigger ventricular system, for instance.

"Brains are like faces. There is lots of individual variability," Dr. Gazzaley said. He concluded: "It's a brain."

So far, so good, just a regular guy.

A FEW WEEKS LATER, after technicians and Dr. Watson in Utah had taken the time to process the rest of the fMRI results, I again visited Dr. Gazzaley's office. This time, he got Dr. Watson on a sleek black speakerphone. It was afternoon, and Dr. Watson was in his office, talking on a cell phone, and cautioned he might get interrupted because he was also watching his seven-year-old son, Nathan, fresh off summer camp for the day.

"He's sitting on the floor behind me, playing video games," Dr. Watson said.

"That's good for his brain," Dr. Gazzaley said. They both laughed.

"They're educational games, mostly."

Dr. Gazzaley had two different images on his two monitors. On the larger screen were several images of a brain, with three key spots lit up in orange. On the smaller monitor was a graph that started high on the left and descended to the right at about a forty-five-degree angle.

The scientists explained that we were looking at measures of Reggie's brain as he attempted to multitask in the MRI machine. The particular activity that Reggie had undergone while his brain was being imaged in the tube is called a dual n-back test. In it, he was trying to juggle two different demands: remembering and responding to audio cues in his headphones and, separately, remembering and responding to visual cues projected into a mirror he could see while lying in the tube. This is represented by the curve on the smaller of Dr. Gazzaley's computer monitors. As the task load increases, performance falls. In that way, Reggie looked like lots of other people Dr. Watson and Dr. Strayer have studied.

So, too, the increased demand on Reggie's brain showed up as a very familiar brain-imaging pattern to Dr. Watson. Three key areas that were lit up in orange were the anterior cingulate cortex (labeled ACC); the dorsolateral prefrontal cortex (DL-PFC), and the prefrontal cortex (PFC). These areas, the scientists explained, are part of the attention network, and the fact that they are lit up reflected an increase in blood flow to those locations. More blood flow means they are taking on a greater load.

That should come as little surprise. Greater demand on attention puts more demand on the attention networks. But one of the things Dr. Watson and others have begun to understand is that people whose performance suffers less—who are, in effect, better "multitaskers"—see a lower load on their attentional networks. The reason is not yet clear, but it might have something to do with how efficiently the brain uses its resources.

Regardless, Reggie's brain, and his performance, was right on par with most test subjects. His attention network took on a predictable load as his performance imploded. As Dr. Gazzaley had put it, Reggie was "just a regular guy."

"There is one thing that surprised me," Dr. Watson said.

It wasn't the imaging, or Reggie's performance on the dual n-back test. It was what happened before—before he got into the imaging machine. Just prior to the test, he'd been asked to take a performance test aimed at establishing his general ability to sustain attention. It's a "baseline" test. How good is Reggie at focusing and dealing with complex problems? Better than most: His score was in the top 25 percent, Dr. Watson said.

"I'd classify Reggie as high in terms of attention," he said. "It really illustrates that even those of us with a lot of attention still have a breaking point."

In other words, Reggie was far from being someone likely to get distracted—behind the wheel or elsewhere. He has a pretty good ability to focus. Nothing superhuman, just solid. And certainly not more of a liability than the typical person. Nevertheless, when his brain was asked to do too much, to multitask, it became overloaded.

"Everybody has a limit. That's the bottom line," Dr. Watson said.

CHAPTER 47

TERRYL

R EGGIE IS A REGULAR guy who rebounded from tragedy.

Terryl rebounded, too. She and her family were happy, the Warner children thriving, the product of a profound generational change in her bloodline.

Jayme and Taylor teamed up and won the state history day contest in 2009, and took ninth place in nationals for a project on Joe Hill, a labor activist.

In 2010 came more remarkable accomplishments. That year, Jayme won first place in the DuPont Challenge, a huge, international science essay competition. She won for her essay titled "Salt: Enhancing Lives One Breath at a Time," which focused on using salt water to help the breathing of young cystic fibrosis patients. Jayme based her research in part on her sister Katie, who is suffering from the disease.

She was the first winner to come from Utah.

But not the last.

The next year, Taylor won first place, and they became the first sibling pair to ever win, let alone back-to-back. His project focused on solving environment problems, notably waste, by using a natural resource: earthworms. The idea had come after Terryl had suggested putting earthworms in the kitchen in a big box to help compost in a natural and effective way.

In 2012, Jayme placed third in the state for National History Day with her paper "Come, Come Ye Saints: Reactions of Early Mormon Pioneers to the Persecutions They Faced."

In 2013, Taylor placed fourth in an international science competition, comparing texting while driving to driving under the influence. "He is now considered a published scientist," Terryl coos.

Taylor graduated early from high school, in 2013, at the age of sixteen, and was the valedictorian at InTech Collegiate High School. It is a small charter school, public, just 140 students, but *U.S. News & World Report* ranked it as the top high school in Utah, and one of the top seven hundred in the United States. In his valedictory speech, Taylor thanked his teachers. He thanked his parents. He thanked Steve Jobs, or, at least, quoted the Apple cofounder:

> *In the first sixteen to eighteen years of our lives, we have lived between 140,160 hours to 157,680 hours. However, between now and the average life expectancy of seventy-five, we will live about five hundred thousand more hours. In mathematical terms, we are barely scratching the surface of our lives at this point. After today, when the caps have been thrown and the parties have ended, we will all begin our journey into the next half million hours of our lives. Whatever we choose to do, we will be well on our way, due to many who influenced us. Steve Jobs once said, "Your time is limited, so don't waste it living someone else's life. Don't let the noise of others' opinions drown out your own inner voice. And most important, have the courage to follow your heart and intuition."*

Before his conclusion, Taylor said:

> *In our future hours of our lives, let us not forget the power of education; the education we have received and the education we*

face for our future. While millions across the world struggle with illiteracy, poverty, violence and war, let us remember that we have the power, determination and creative minds to help fight and eliminate those social problems. I hope that in our future hours, we will find time to remember our responsibilities to others and resolve to give something back to our society.

For his part, his heart was set on becoming a neurosurgeon. As for Jayme, she was headed to her mission in the Philippines, and then she, too, wanted to go to medical school.

The younger girls, Allyssa and Katie, were spending the summer of 2013 doing opera and musicals, Allyssa performing in *Fiddler on the Roof* and Katie appearing in *Joseph and the Amazing Technicolor Dreamcoat*. Terryl did the driving to and fro, and waited by the stage as her girls practiced. And they performed with the Utah Festival Opera.

Allyssa, twelve, a huge fan of music, composed an original piece about Joe Hill, the same labor activist on whom Taylor and Jayme had done their project a few years earlier. Allyssa and her partner came in first in the state's history day in the performance category.

OF COURSE, THERE WERE ghosts from the past. Terryl struggled with her mother, Kathie, who now lived in Logan. They would occasionally clash. Taylor didn't think his grandma liked him much. Terryl would get frustrated that her mother didn't seem to want anything to do with the past, wouldn't so much as brook a conversation about it. Terryl told her mother she was talking about her childhood for a book. Terryl says Kathie thought it was a bad idea, and she herself has declined to be interviewed.

"She does not understand why we have to tell anybody. For her, nobody has the right to know," Terryl says, in the rare raised voice.

"Here we are, years later, *years later*, and the idea is, let's not let anybody know—let's pretend we did not have secrets destroy parts of our life. Yes, yes, I did. My mother's idea is we don't talk about things because we don't want anybody to know. I get really angry."

Terryl says: "I'm sick and tired of shoving things under the carpet. People have a duty to stand up," and not just within their families. "In society, people don't want to stand up."

She and her brother Michael corresponded some by email. He has professed to be happy. He also has mentioned to Terryl in email that he had a heart attack in 2005 and that it caused his spirit to leave his body, and travel and get wisdom. He writes of conspiracies and satanic plots. He once wrote and told her that all the punishment and whippings doled out by Danny had, for all their horror, one positive side effect. "You have to give [him] some credit though . . . he made us both unusually strong. I do not think I know of any woman in this world that actually is as strong as you, and that has a lot to do with the treatment ol' BLD gave us."

Terryl's younger brother Mitchell feels these accounts are unfair. His dad, he reiterates, is his hero. How to explain how his memories differ from those of his siblings? Family friend Donna Simpson says Mitchell, who came along much later than Terryl and Michael, got different treatment by virtue of being Danny's biological offspring. Mitchell thinks his father and Kathie were a volatile mix and that his dad might've been different with Michael and Terryl before the parents split. But Mitchell offers another theory, too, about how his experience differed from Terryl's. It has to do with how people view the world differently. Two people can be in the same car crash, he says, and describe it very differently.

BUT, ON THE WHOLE, life for the Warners was about activity, family, and it was about service—church every Sunday—and kindness, the things that Terryl had pleaded for in her diary as a little girl. Earlier in the year, one of Jayme's childhood friends, a boy, came out to her as gay. He was being shunned and bullied, even turned away by his family, which was something that could happen in a small, highly religious community. Terryl told Jayme to be his friend, to listen and not judge, to help him be himself and feel comfortable with who he is. Jayme says "When Jesus said, 'Love everybody,' he didn't make any exceptions to that rule."

Not that Jayme needed to hear that; she says when her friend came out to her, it didn't even dawn on her that it was an issue.

"I told Alan, 'I think we've done something right for her to embrace this young man,'" Terryl says. "She doesn't care if he's gay or straight. That isn't what life is about."

Terryl says the boy later told her: "I just want to thank you, because your daughter saved me from doing some drastic things." He also planned to go on a mission, and wants to start a family.

Terryl seemed to have developed, and passed on, a deep moral authority, despite her tragic childhood, or maybe because of it, that seemed to have more depth and breadth than the institutions and conventional leanings around her.

The Warners were raising joyful, happy, family-centric children. They kept at arm's length the multimedia that Terryl feared would foil their engagement with one another, their studies, the real world. She saw this as crucial. Too much media would be numbing, dulling, cheat her children of an engaged life, the kind she'd always craved as a child. One thing was for certain: the life of the Warners, and Terryl's adulthood, was a far cry from the path that might have been.

What could they teach us? What could Reggie teach us?

CHAPTER 48

REDEMPTION

THERE ARE TWO RULES.

In the public-policy debate around distracted driving, safety officials have drawn on those two seemingly tried-and-true rules used to reform drunken-driving culture and seat belt adherence: strong laws and enforcement, coupled with public education about both.

That's how society changes.

But how do individuals change?

How do they come to terms with something tragic, dissonant, dangerous, contrary to their long-term interests, their safety, the interests of their children? How do they heal, themselves and others?

Here, after talking with many experts, filtering many views, it seems there is one rule. It is simple, yet not nearly so easy to achieve.

Tell the truth.

To yourself, foremost, but also to your friends and family, to the world around you. This is not to say that everyone has the same truth, or that there is a universal truth. There are personal truths, big and small. Do I have an addiction? Am I acting in a way that is consistent with what I preach? Am I abusive? Am I a victim? Where are the disconnects in my life? Where is the deception, or self-deception?

Honesty can be hard, obviously. That's in part because it can be so hard to even see the points of conflict or dissonance. We have powerful defense mechanisms—personal prejudice, shame, societal messages, and, yes, technology. But all of those, once recognized, can be overcome. And once someone finds the capacity to tell the truth, finding the truth, and expressing it, becomes much easier.

REGGIE COULDN'T STAND THE idea of letting his family down. He couldn't miss going on a mission. So he lied about Cammi.

When he got into the wreck, he thought he hydroplaned. Then his mother told him that his brother had instructed him not to say anything else to the police. And then he saw his dad's terror that he might go to jail. His lawyer said not to say anything. He'd hydroplaned, right? That's what he'd told them. He didn't remember texting. He was just driving.

If he pleaded guilty, or admitted something he and others had convinced him he hadn't done, or couldn't be proved, he'd again put his mission in jeopardy, he'd risk letting down his family.

WHEN DON LINTON WAS a boy, a time in one's life when innocence and truth should go hand in hand, he was a walking, living zombie of a lie. How could it be any other way?

The center of life was the Church, the beacon for all good, but a member in good standing had raped him repeatedly. There was no one to tell. Maybe he couldn't even see the difference between truth and lies anymore. He even wondered who was at fault: him or the monster?

He couldn't tell the Church. He had trouble even telling himself; he got straight A's, he put on a face. And so it stayed trapped inside him, poisoning him.

In school, he says, "I lived a facade.

"I wanted to be liked so I made myself likable. I had the outward appearance of somebody who should be loved. Everybody thought I was smart. I was popular. But when I left school, I left the ability to impress. All of a sudden, the crutch was gone."

He spiraled and spiraled. He took heavy medications to pave over the dissonance. Anything but the truth.

TERRYL WAS DIFFERENT, PERHAPS the exception that proves the rule. Somehow, blessed by some spark, she started telling the truth, or trying to, at a very young age, when all the influences were telling her to go a different way. It wasn't just that she wrote her truth in her diary. She poured out Danny's booze, and she risked beatings to do it. But it was more complex than that. She often told an outward lie, willing herself to be peppy in church, at school, using positive affirmation, which turns out not to be at all the same thing as telling the truth.

Positive affirmation is a temporary fix, a surface measure. Reggie kept telling himself he could get through it, trying to man up, and endure.

"You have to tell your emotional truth," says Susan Forward, a therapist and the best-selling author of *Toxic Parents*. This can mean being honest about your own behaviors or emotions, or confronting truths with people who harmed you, as Terryl did, and Linton did, and still struggles to do.

"You have to confront," she explains. Yourself or others. "You have to say: 'This is what you did to me, and this is how I felt at the time.'

"You may be damaged, but you can come out of the sewer and you can come out okay," Forward says. The value isn't merely for the person who was hurt, or who hurt others. In order to heal a family, the next generation, or society, she argues, "It's vital to make the connecting thread between the past and the present."

To Forward, Terryl is a prime example. "Her kids are the next generation, and she had the courage and awareness to change and say: 'My kids will not go through what I went through.'"

Of Terryl, Forward says: "She is a hero."

That said, Forward believes it is naive to think that all people can survive trauma, as great as those experienced by, say, Terryl and Linton—or, in a different way, Reggie.

They were just lucky to be wired to endure and overcome. "Why is one person who didn't have such a bad childhood so ill, and why is another person who had such a horrendous childhood able to function at such a relatively high level? My conclusion is: A lot of it has to do with the way we're wired," Forward says.

TERRYL GAVE A LOT of credit to her religion, support from the Church, her abiding belief. Reggie feels the same way. In the end, religion was a great aid to both, but not a substitute for the lessons learned from hard experience. In the end, both recovered a sense of moral authority from the ashes of those experiences, which allowed them to challenge the world in a way that even those with the deepest abiding beliefs might not have been able to do.

As to the role of religion in healing, the experts I talked with feel divided. Forward, who has run fifteen thousand therapy groups, says: "A lot of people turn to religion for the strength they don't have inside them. They turn to God as the ultimate therapist. For some people, that is comforting. But I don't know what they do when it lets them down."

On the other hand, Marc Galanter, who founded the Division of Alcoholism and Drug Abuse at New York University, has done extensive research into the power of Alcholics Anonymous. He says that a crucial aspect of breaking the bonds of addiction and finding inner peace owes to "a commitment to a higher power and spiritual awakening."

He believes that, over time, neuroscience will be able to validate some of the benefits of such an experience. It's not the same thing as religion being true—that is a different question—but more speaks to the value of providing something people truly believe, at least in the case of addiction. "These cultic movements and AA can achieve dramatic and rapid transformation," he says, speaking very broadly of religious movements.

Simplistically put, religion works for some. Not for others.

RELIGION DID NOT HELP Linton. In a way, it was a hindrance, at least the institution of the Church.

In the end, Linton feels he could only tell himself the truth because he met someone who would help him hear it. When introduced to his eventual wife, he says, she saw something in him, on him—the suffering. He began to allude to what was happening, to that idea that something had happened. And then, he says, as the truth oozed out, she was there to help him own it.

"It takes love," he says.

"It takes somebody who understands why you're being so flaky, why you're worried about leaving your kids with the Boy Scouts, why you're sobbing over your sister's death twenty years after the fact. It takes someone who can love you even though you look like a living skeleton," he says, referring to his distraught state when he first met his wife. He credits her with saving him. "It takes this kind of unconditional love—somebody seeing this value in you, even when they know the truth."

Linton sees considerable similarities between his own capacity to rebound and Reggie's ability. "This kid has people who love him unconditionally. They understand what he did was a horrible, stupid, act," he observes. "But his parents absolutely adore him and realize that Reggie's not a bad kid."

Linton goes further in drawing parallels among himself and Terryl and Reggie. "Throughout the world, a lot of what's needed is to just bring public awareness to what is real—whether sex abuse, or childhood abuse, or texting on the highway. It's easy for me to compare it," he says. "Even though with men it's different—the state of mind is different—

the result is not that much different. You destroy a life, and even if you don't destroy a life, you maim the person."

Prosecutors, he says, lose cases all the time around childhood abuse, around issues where it is hard for people to acknowledge the reality of the terrible things people can do to one another. "There's still a lot of denial out there."

There's another connection between Reggie and Terryl: addiction. For Danny, it had been booze. In Reggie's case, it's impossible to conclusively say he was addicted to technology. But he did find the activity almost irresistible, so much so he did it behind the wheel, maybe to fill a social void, and feel connected. For many others, particularly since technology has gotten more powerful, the signs of compulsion are even stronger. If the electronic gadgets are not used in fair moderation, they can take over a person, maybe not to the extent that drugs overtook Danny, but certainly in ways that can change who they are, the kind of parent or friend they become, how they learn, how they attend to the world.

TELLING THE TRUTH TO others is often the easier part, the confrontation. The harder part usually is telling the truth to yourself. That owes partly to another of our great survival mechanisms that can be problematic when relied upon in excess. It is our capacity for self-deception.

By way of example, if people really thought about how dangerous it is to drive a car (it's the most dangerous thing they do on a regular basis, given the likelihood of a crash and a fatality), they might never get on the road, unless they put at bay some of the realities.

Dr. Atchley says that, at any given moment, a person really only sees a tiny fraction of space, but we allow ourselves to believe we see a wider scope than we actually do. "It's a grand illusion," he says.

"So much of our brain is built to shield us from knowledge that we are far less in control than we think we are," he adds. "One of the brain's most amazing abilities is for self-deception. So when there's a disconnect between attitudes and behaviors, the brain changes attitude rather than changing behaviors."

That self-delusion becomes more powerful and perverted in the case of addiction, argues Dr. Greenfield.

"It's the nature of addiction. It's self-cauterizing, self-soothing, dopaminergically numbing. There's no reason to change if it's working—and it *is* working. People do drugs—and I include the Internet—because they work. They numb you," he says. And they keep working until some external force causes you to awaken to a problem—a parent, a coworker, a spouse, or something tragic, like a fatal wreck. "When it comes to addictions, whether alcohol or technology or the Internet, it's almost one hundred percent motivated by an external force."

"There has to be a walk in front of a mirror, even an accidental walk," he says. For instance, he explains, you could be gaining weight "and walk past the mirror and get a glimpse of yourself even if you're not looking at it, and that starts a cascade of self-analysis."

In the case of technology, phone use or Internet addiction, he says he hears lots of versions of a story in which someone loses the phone for a day, or the power goes out. "They are now juxtaposing their typical day against this new feeling, what it feels like to not have cortisol going through the roof," he says. "That's when they hold the mirror up to themselves."

THERE ARE FAR MORE differences than similarities between what Reggie experienced and what Terryl endured in her terrible childhood, or what Linton suffered at the hands of an abuser. But they all went on to contribute, to fight for causes, to compensate and grow through a kind of service.

To Forward, this owes to their abilities to have confronted the truth, expressed it, become genuine—in a way that isn't possible through platitude or mere positive affirmation. "I genuinely believe that people who overcome these things have a special place, a core courage, a spirit of integrity that no one else can get to," she observes. "I can't explain it clinically. I just understand it in my gut.

"Wounds are an enormous sense of wisdom, if you use them."

NOT EVERYONE HAS A trauma to confront. But many people have half-buried truths, or small lies, or disconnects, little bits of dissonance in how they live their lives. Are we attending to the world—our families and partners, coworkers, friends, jobs, and ourselves—in a way that is consistent with our stated values?

That is how all of this connects to technology, and how we use it in our daily lives. Reggie's story, in the end, isn't just warning us not to use technology behind the wheel. It is screaming out to us as a society: pay attention. Know the risk technology poses in your life, and tame it.

Sure, there are lots of little rules, actions that a person might do to limit technology, like turning off the function on a computer or phone that flashes an instant update every time there is a new email or Facebook status update. Or by putting the phone in the trunk of the car when you drive, thereby saving yourself from fending off the impulse that assaults the bottom-up attention, which sends a desperate signal to the frontal lobe and interrupts the driving task.

The journey that travels from Donders to Helmholtz, through Broadbent and Treisman, to Posner and Strayer, tells us a few basic things: That the brain is limited, lacks bottomless capacity, and isn't particularly fast relative to computer technology. The technological advances defined by laws of Moore and Metcalfe got married and long since overtook what our brains can handle. And we know now that their union also presents the challenge of addiction, or, at least, serious compulsion. Even allowing for the fact that we can know these devices challenge our limitations, sometimes they are so compelling we can't stop using them.

To ignore this science is to engage in self-deception, to tell ourselves a lie.

And it gets supported by neuroscience.

REMEMBER THE CHOCOLATE CAKE study, the one that looked at the nutrition choices made by someone asked to remember numbers? It showed that a person's decision making is impinged by something as simple as a modest memory task.

The choices someone makes—and the clarity of their mind—directly relate to how much information the person is trying to juggle. There are a bunch of studies that support, directly and indirectly, the significance of not being overloaded with information.

A remarkable bit of research was done with veterans returning from Iraq and Afghanistan. They were excitable, some suffering from post-traumatic stress disorder. The research, led by Emma Seppala, the associate director of the Center for Compassion and Altruism Research and Education at Stanford University, found that the veterans saw a significant decrease in their startle response when they did deep breathing. Put another way: They were less impulsive and responsive to external stimulation. Dr. Seppala says the research could be extended to help people become less responsive to their devices, to exhibit more top-down control of their environment, rather than have it drive them.

Another study from the University of California at San Diego found that creativity is enhanced when people let their minds wander, or when they do the most menial of tasks. Other research reinforced the risks of overload to decision making and learning, or, looking at those studies from a different perspective, proved the clear value of taking breaks from constant stimulation, whether provided by technology or some other high-fidelity experience.

A study at the University of Michigan compared people's ability to learn and remember information under two different conditions. In one condition, the subjects walked in a dense urban environment and then tried to learn and remember information. In the other condition, the subjects walked in nature. Those subjects, the ones who walked in nature, did a better job at learning.

To researchers, the study suggests the restorative value of nature but for a key reason: In nature, people are processing less information—not zero information (they are taking in the world around them)—but less dense, intense information. Even the background noise of a city can impact the ability to learn and remember, and can deplete neurological resources.

In another study, researchers at the University of San Francisco appear to have found more evidence of the cost of constantly consuming information. The USF scientists discovered that when a rat (their test subject) has a new experience, say, exploring a new environment, it stimulates new patterns of brain activity. But, the researchers concluded, the rats don't appear to encode that experience as learning and memory unless they have some downtime after the experience. In other words, if the rat continues to be awash in new experiences, and doesn't give the brain a break, it doesn't process the new information.

The sum of these studies, and others along the same lines, offer a pretty clear road map to reaching a state of mind that allows for good decision making and better awareness of one's life—the precursors to balance and happiness.

Take breaks from stimulation.

Put another way: Turn off the device for a sustained period—whether hours or days. And then, and here's the really tough part, don't fill the void left by the absence of stimulation with some other nonstop stimulation. That can be hard to do, the experts told me, when people are so accustomed to the constant stream of pings and external noises. A veritable panic can ensue—*What do I do with this void?* But it's there, when things die down, that the learning and memory get strong but, more than that, that the greater powers of decision making come into play. When a person is clearheaded, the frontal lobe becomes freed of the humming and buzz of external pressure. That's when you can decide what steps are best, what actions are wisest.

"Don't put down the phone just to get preoccupied by something else," Soren Gordhamer told me for a story I did for the *New York Times*. He's the organizer of Wisdom 2.0, a growing movement in the Bay Area aimed at helping people find balance in the modern world. "If you're not careful, the same orientation you have toward your device, you'll put toward something else."

The neuroscience, while providing support for the idea of taking breaks, also was providing the seeds of another solution to distraction: a high-tech solution—with Dr. Strayer and Dr. Gazzaley, among others, at the forefront as well.

CHAPTER 49

THE NEUROSCIENTISTS

D R. STRAYER WAS INVITED in 2010 to the Transport Research Lab, a swanky, glass-paneled center in Workingham, about thirty miles west of Central London. He was there to talk about a group of individuals whom he and Jason Watson were calling the "supertaskers." It was this type of person whom Dr. Strayer had identified several years earlier, by accident, when he and his student discovered a study subject who seemed to be able to do two things at once without much cost to performance.

In the interim, the pair had identified a handful of such people. There weren't many. But they did seem to exhibit some profound characteristics, not just behaviorally, but neurologically. When the researchers scanned the subjects' brains, they discovered that this tiny fraction of people seemed to have less activity going on in key portions of the frontal lobe. "Their brains were less metabolically active," as Dr. Strayer explains it in slightly more scientific terms. In other words: When the people were doing two tasks, "their brains weren't working as hard to do it."

Dr. Strayer and others developed a hypothesis that these people "are extremely good at neural efficiency." Meaning: They don't need as much of their brains to process information and therefore might be able to add on more tasks without overloading the system.

The idea at the research lab in Workingham was something of a gimmick, a "media event" designed to show some of the emerging research. Scientists in Britain had found a few people who seemed particularly adept at multitasking—potential supertaskers—for Dr. Strayer to put through the paces. Most did okay, but there was one woman whose capacities blew Dr. Strayer away. She was in her late twenties, a high-level speed cyclist, Olympic caliber. To establish her cognitive baseline that day, the researchers had her do some math problems and memory challenges. She missed just one question.

They put her in the high-fidelity driving simulator. Her score on the cognitive test rose, what little it could. She got a perfect score, as she was navigating intense roadways with ease. "I'd never seen it before," Dr. Strayer said, adding that the challenge of the dual tasks "are designed to destroy most people."

After years of focusing on distraction, Dr. Strayer was fascinated by this group of people. He started collecting their DNA. What was it about these people? He ran a test on one of their neurotransmitters that regulates dopamine, which in turn is associated with attention. He came up empty. Nothing was easily presenting itself. The sample size was too small, the haystack too big, and the needle invisible.

But the concept of how their brains work held promise for Dr. Strayer. "Maybe if we can understand something about extraordinary behavior, we can understand how ordinary people can do it."

And in this analysis, Dr. Strayer offers another key thought: Ordinary people aren't supertaskers. Far from it. Ordinary people do better at a task when they focus on it, not try to juggle it with another task. Ordinary people aren't like the woman in the experiment outside London. Ordinary people are like Reggie.

The other thing Dr. Strayer's work underscores is how the neuroscience community was looking for answers not just to how attention works, or how it can be improved through behavioral measures (deep breathing), but by tinkering with the brain itself. Could this be the answer? We don't change the technology, we change us?

A TWENTY-TWO-YEAR-OLD STUDY SUBJECT sits at a table in dim light. In her hands are an iPad. On her head is a cap with thirty-two wires coming from it. They are attached to electrodes, which measure the electrical activity in her brain.

A few feet away, through a wall and over a closed-circuit camera, Dr. Gazzaley and a volunteer assistant named Jesse watch the study subject, a teacher who lives in the East Bay, just across the Bay Bridge from San Francisco.

This experiment is part of Dr. Gazzaley's $10 million new lab, located in the basement of the Sandler Neurosciences Center, five floors below where the neuroscientist has his office. Generally, academic labs sound much fancier than they are; often they are windowless rooms, cubbyholes, and former closets dug out and jerry-rigged by cash-strapped researchers, but this one is different.

Dr. Gazzaley, though he is fighting for every dollar, has managed to oversee the construction of something sleek, more in his image. In two adjoining rooms, there are MRI machines. Here, in the control room, three monitors on the wall can show what is happening in each of the rooms.

On one of the computers on the desk, a bunch of data begins to pour in. It is showing what's happening inside the subject's brain, the electrical signals. It is not decipherable in the moment, but eventually it will offer a picture of what is happening inside the study subject's brain as she's asked to distribute her focus more widely.

On the screen of her iPad, she plays a relatively primitive video game that entails identifying targets that at first appear in the middle of the screen but get farther out on the periphery. What Dr. Gazzaley and his team want to do is not merely understand how the brain works when confronted with a demand to widen attention. They want to then put that knowledge to work to try to increase the brain's capacity to broaden the limits of attention.

"We don't know if it's pushable," he says. He's smiling, almost sheepish, as if to say: We're going for it, but who knows.

But if they're pushing ideas here—trying things that just may not work—this is hardly mad science. In fact, in the summer of 2013, not long after this experiment got under way, Dr. Gazzaley and his team scored a huge scientific coup. They found out the journal *Nature* would publish their paper that described how they could train the brain to improve and sustain attention using a video game.

The research, which took place over four years and cost $300,000, entailed training 174 adult subjects to play a targeted video game. The game was called NeuroRacer, a three-dimensional driving game that asked the subjects to drive while simultaneously doing perceptual tasks, like recognizing objects around them.

"It is clear that multitasking behavior has become ubiquitous in today's technologically dense world, and substantial evidence has accrued regarding multitasking difficulties and cognitive control deficits in our aging population," reads the introduction to the *Nature* paper. Translation: As people get older, they have more and more trouble keeping up with switching tasks and focusing under demanding circumstances.

As the subjects used the video game, the researchers studied their performance with behavioral tests, and used eye scanners and captured the electrical activity in their brains with EEG. As their research wore on, Dr. Gazzaley and his peers discovered something they found remarkable: Older subjects, age sixty and above, showed increased ability to focus. What was remarkable was that this increased capacity to sustain attention came not within the game, but *after* they'd finished, weeks after.

"Critically, this training resulted in performance benefits that extended to untrained cognitive control abilities (enhanced sustained attention and working memory)," the paper concluded. The EEG research found intensification over time of theta waves in the prefrontal cortex, thought to be a measure of sustained attention. That change predicts a "boost in sustained attention and preservation of multitasking improvement six months later."

The findings, the paper said, "provide the first evidence . . . of how a custom-designed video game can be used to assess cognitive abilities across the lifespan, evaluate underlying neural mechanisms, and serve as a powerful tool for cognitive enhancement."

"The research shows you can take older people who aren't functioning well and make them cognitively younger through this training," Earl Miller, a neuroscientist at the Massachusetts Institute of Technology told me. "It's a very big deal."

Indeed, the research earned kudos from a range of attention scientists. Among them is Daphne Bavelier, a neuroscientist at the University of Rochester, who has been one of the pioneers in trying to understand how technology could be used to expand and improve attention. In her case, she was able to show that young people playing off-the-shelf, shoot-'em-up video games showed an improved capacity to sustain attention, even after the game was over. Neither she, nor Dr. Gazzaley, believed their findings should be taken to mean that the simple act of playing a video game improves the brain, or, specifically, the attention networks. Rather, she says the technology—coupled with brain imaging—began to offer hope that real therapies could be developed for attention-related issues.

She said Dr. Gazzaley had achieved a big leap by showing that a scientifically developed game can reverse or "rewire" some of the effects of aging—not just in the game, but months after the people stopped playing. "He got it to transfer to other tasks, which is really, really hard to do," she says.

For her, like Dr. Gazzaley, the applications are potentially widespread—someday we may be able to treat attention deficit disorder, learning disabilities, and attention-related memory loss without drugs, which can have side effects. And these technological therapies would have to be constructed, she said, so that they don't invite their own side effects, like, say, overbuilding one neural network. "We know the brain is rewirable, the question is to rewire it properly.

"We're in the primitive age of brain training."

Echoing Dr. Gazzaley, she said she was thrilled at the idea of turning the tables on the technology, making it a slave in new ways rather than have it challenge the brain.

"Technology is here to stay. Like any revolution, there are pluses and minuses. We need to harness that power to our advantage."

This has been, all along, what most excites Dr. Gazzaley—not just the hunt to understand attention, and technology's impact on it, but the effort to expand it. Take, for instance, the room in his new lab next to the control center where he's been examining the brain of the teacher wearing the EEG cap. In this other room, there are the latest video game machines, like Microsoft's Kinect video game console, which uses motion-capture technology to allow a person to interact with the action on the screen. It's hooked up to a forty-six-inch Samsung monitor, its empty cardboard box still sitting in the room. There's ultra-modern science technology here, too, including an EEG cap that is wireless, meaning that it can send signals without being hooked up to wires. Dr. Gazzaley's idea is to marry the consumer technology with the research tools to ultimately try to create ways for people to improve their attention, focus, and learning ability from the comfort of their own homes—using technology.

"I want to help rebuild these circuits," he says. "We can sculpt and shape the neural circuits to refine the brain. We can drastically improve people's abilities.

"This is my dream."

CHAPTER 50

REGGIE

IN 2012, REGGIE STARTED dating a woman named Britney, who was twenty-two at the time, three years younger than he. They'd been friends for a few years while they both worked for the Utah Jazz, the basketball team. By the fall, they were going out regularly—movies, bowling, miniature golf.

She knew about the accident and seemed to accept him. She had her own struggles growing up, and they felt comfortable in each other's presence, knowing they had imperfections.

Reggie was living in an apartment complex in a busy part of Salt Lake City, surrounded by commercial strips, chain restaurants, and retailers. His second-floor, one-bedroom apartment was decorated in the way of a twenty-something male who worked a lot; frozen food in the freezer, a twin bed with a comforter he periodically took home to his mom to wash, a bookshelf with some schoolbooks and movies and video games he didn't have time to play. In the small living room, there was a hand-me-down love seat and a chair, from a relative, and the one thing in the place that had a feel of personal touch: a poster of Muhammad Ali standing over a defeated Sonny Liston.

Reggie loved Ali's courage and tenacity. Reggie even played around with boxing. In the fall of 2012, he fought an amateur bout in Las Vegas, which drove Mary Jane nuts. Her little guy was going to put his precious head in harm's way?

But Reggie felt like he was trying to do some new things, live life in a more full way. He took piano lessons and had a public recital.

And work was going great. Reggie had landed a job as a manager at a basketball and fitness gym just south of Salt Lake City proper. It was a huge facility, a place for the casual workout experience, but also a facility where professional athletes—and high school and college ones—could work out, play ball, lift weights. And Reggie was running the whole show, with at least fifteen people reporting to him. He was putting to work his tenacity and work ethic, the way he always gave to his teammates growing up, and blossoming into a leader.

REGGIE DECIDED TO TAKE a chance. On September 16, 2012, nearly six years from the day of the accident, he told Britney he had a surprise of sorts for her.

"I want to take you somewhere," he told her.

They climbed into his Mazda. He was nervous, to say the least. They drove up I-15, passing the Wasatch, the route he'd taken when he returned from his first mission. Then they turned toward Logan, passing Chocolate Pass. By now, he'd managed the courage to pass by the accident site a few times. Now, he actually stopped.

It was around ten a.m., the sun still coming up, the skies clear, not like the morning it happened. When they got there, Reggie, for the first time, walked through what he remembered had happened, and told Britney the details he remembered, showed her where his car wound up. He was crying, and she was not. "She was strong for me," he said.

After a while, they drove to the cemetery where Keith O'Dell was buried. Reggie and Britney sat down near the headstone and didn't talk.

A few days later, Reggie felt like he might be turning a corner. "I think it was some closure for myself," he says, but then he pauses. "I don't know if that was a turning point, but I'm extremely grateful that I did it."

The reality was that each time Reggie seemed to feel he might hit a turning point, he'd slide back. He'd have dark periods. He'd feel depressed. He wasn't sure why he was turning back; every time he did a presentation, as his dad feared, it would bring back the terror and shame. But when he didn't do those things, he felt guilty.

How could he stop testifying to the risks of texting and driving when someone out there might still do it, and take a life?

WHEN IT CAME TO Reggie, Terryl was among those who felt Reggie had done enough. Only Reggie seemed to wonder.

Jackie, Megan, and even, to some extent, Leila were impressed by what he'd done. Jackie had found a new partner. Stephanie and Cassidy were doing well in school. In fact, in the fall of 2013, Cassidy skipped ahead a grade—bypassing fifth grade and going right to sixth, soaring on the strength of her math skills (not unlike her father).

Around that time, the girls and Jackie got together with Reggie and Terryl. Part of the reason was that Stephanie and Cassidy had decided to team up to enter the history fair. They chose the topic of "rights and responsibilities," and planned to focus on the right to drive but the responsibility to do so without using a phone or texting. And they'd focus on the accident that took their dad.

They had lunch and then went to a park where the girls, including Terryl's daughter Allyssa, played on the jungle gym. The adults chatted. Jackie told Reggie about the evolution of her feelings, about how she'd been angry, and had just wanted an apology from him or some flowers after the accident. That would've meant a lot. But she said she'd forgiven him. Soon Stephanie came over and said she wanted Reggie to know she was doing just fine.

"Reggie could see that we're okay and everybody has choices to make and our choice was to move forward," Jackie says.

She told Reggie that Jim loved life and that he would want Reggie to do the same—live life to the fullest.

MEGAN O'DELL HAD DIVORCED her first husband and gotten remarried, but still couldn't find a job situation she liked. She was drinking some, maybe more than some, spending hours playing video games. She and Reggie would talk regularly.

He really wanted to help reach out to her. And Terryl felt deeply for her, too. "I didn't have a father at my wedding," Terryl says, sympathizing. For her part, Megan wouldn't blame Reggie for the challenges in her life.

Leila, when at her most open, mourned not just the death of Keith, the love of her life, but Keith as a conduit to the world. Leila was a loner. He gave her connection. In other words, it wasn't just Keith she missed. It was the world.

A few months before Reggie got together with the Furfaros, in the spring of 2013, Leila expressed what for nearly seven years had been an impossible thought. She wondered if she might meet someone new. It was about needing and wanting to connect with the world. She even wondered if she might try an Internet dating site. For the most part, the idea made her feel squeamish, as it does for lots of people. But she was distantly open to the idea that the Internet, this wondrous technology, could help a loner take some baby steps to connect, even if only with a new friend. The idea wasn't to replace Keith—not ever—just to establish some new ties with the world.

She still hadn't met Reggie, or spoken to him.

"I don't want to talk to him. What's he going to tell me, that he's sorry? That doesn't do me any good. How is that going to change my life?"

I asked if it might change his life. And she said: "Well, right now, I'm not too worried about him."

In the case of Reggie, his back-and-forth relationship with happiness took a steep dive in the summer of 2013 when AT&T, the mobile phone company, decided to make a video about the accident involving Reggie and Jim and Keith. They hired a famous movie director named Werner Herzog.

Megan was excited by the idea. Terryl didn't see why not. Reggie was scared but didn't want to let anyone down. He agreed to join a film crew back at the site of the accident. They gathered in early June. At the scene, Reggie imploded.

"He was shaking, crying, sobbing. He couldn't function. He could not talk," Terryl says. "He grabbed me and held me and said: 'I'm so sorry, I'm so sorry, I'm so sorry.'"

She held him.

After a while, she said: "Reggie, you're in no position to drive."

"I feel protective of him now," Terryl says of Reggie. "It's a mother type thing."

Leila showed up over the next few days when the film crew was interviewing people. Megan, Terryl says, was acting kind of crass, burping and being a bit strange. Leila seemed embarrassed by her daughter's behavior, but otherwise "happy."

"She smiled. I was really amazed. She asked me about Reggie, and she did not refer to him as 'that Shaw kid.'"

"How's he doing?" Leila asked.

"He's not a bad guy," Terryl told her. "He's done everything. He will never be able to give back what he took from you, but he's trying so hard."

Megan seemed to agree with the assessment that Reggie was a great guy. She told Terryl: "That's the kind of guy I hope I could see myself winding up with."

Reggie just couldn't view himself that way. The AT&T thing really crushed him, the wound torn open again. "I don't know how I could do something so terrible to people.

"I made the decision, so the rest of my life I'll have made that decision. That's never going away. I can tell billions and billions of people and that does not change the fact that those two men were killed.

"I can try to justify it, or make excuses, and try to explain. But when it comes down to it, I shouldn't have done what I did. It happened. It shouldn't have happened. I shouldn't have done it."

A FEW WEEKS LATER, he was back on the stump again, this time talking to the rookie camp at the NBA. He did his thing, and he was feeling very low, the strain and grief pouring out of him.

A psychologist who works for the NBA took notice and pulled Reggie aside, and the two men talked for two hours. The psychologist gave Reggie an analogy relating to airline travel.

"You know how when the safety video comes on, they tell you to put the air mask on yourself first before you put it on your children?"

"Yeah," Reggie said.

"They do that because you have to save yourself before you can save other people."

The analogy stuck with Reggie, and he kept wrestling with it. "I don't have my mask on and I'm trying to put everyone else's mask on," he said.

He was thinking of listening to the psychologist's advice and taking some time to get his own mask on. There was another experience that reinforced that idea. A few weekends later, there was a tournament at his basketball gym. Reggie organized it, managed it, and he put out some materials from AT&T about the risks of texting and driving, along with materials he had created himself—T-shirts, bumper stickers, and videos.

It felt great making that part of the event, "and I didn't have to speak one time."

It was a middle ground, giving without opening the wound. But, despite that, Reggie still wrestled with the guilt; he didn't want to give up the grief too easily. A few days later, he was back in a funk, vacillating again; he didn't want to act like he hadn't made a terrible choice. So what if more testimony meant he had to suffer a little more.

"I've got a job, and an apartment, a car, a good family, a girlfriend. I'm doing those things to help other people, and I'm doing all right. It's not like I've got no home, and nothing to eat, or no way to provide for myself. I'm getting by.

"If you can save someone else's life and you don't do it because it hurts you, how selfish is that?" he asked. For Reggie, it seemed there would yet be more wrestling, and, ultimately, a question on where he should put his attention—on himself or everyone else.

"What if I miss all those opportunities to help because I'm so focused on myself?"

EPILOGUE

IN 2013, NOT LONG after the Google glasses made an early appearance at the Gazzloft, a picture began circulating on the Internet. It showed one of the first products of the camera app for the Google glasses. That first picture was taken by a driver wearing the device—purportedly one of the official testers of the fledgling product—who was using the wearable computer while cruising down a huge San Francisco hill.

One of the things public safety advocates fear in the distracted-driving conversation is the glorification of multitasking.

It has an insidious, if not overt, quality of reinforcing the idea that it is uncool, foolish, to not be connected all the time. It's a message that comes not just from the technology industry but from cafés, sports stadiums (log on during the event), and airlines (offering constant connectivity as the great in-flight perk). There's a basic understanding in the culture and in our everyday language that it's better to be connected all the time. That it's cool. And that the reverse—being disconnected—is worse.

Take just a few examples, like the ad campaign that ran in airports in 2013 promoting the Motorola Droid Razr Maxx HD phone, an image that featured hipsters staring at their screens and a tagline:

"For those who can't put it down. 32 hours of battery life and a brilliant HD screen." Or there's the implicit, or explicit, value of multitasking, as evinced by a popular advertisement from AT&T in which a guy in a suit and tie sits at a table with a group of four toddlers.

"What's better," the man asks the kids, "doing two things at once or just one?"

"Two!!" the kids shout.

"You're sure?" the man asks.

One of the boys says: "It's two times as awesome."

The tagline of the ad: "It's not complicated. Doing two things at once is better, and only AT&T's network lets you talk and surf on your iPhone 5."

It's not complicated—you'd have to be as small-brained as a toddler to not get it, duh—yet, as we've seen, it's also not technically possible from a neurological perspective. You can't do two things at once, but you can switch back and forth, task switch. The ads from these technology companies do not specifically advocate doing two things at once in the car. In fact, AT&T, after years of fighting against laws aimed at banning cell phone use by drivers, has an active and vibrant campaign against texting and driving.

However, safety advocates feel some of these companies are walking a fine line by glorifying the value of constant connection while vilifying texting and driving. "People certainly are getting mixed messages," says Barbara Harsha, the longtime traffic safety advocate from the Governors Highway Safety Association.

This kind of marketing that celebrated constant connectivity drove Reggie nuts. There was one TV commercial he saw in the summer of 2012 in which a young woman was getting into her car, bragging about how her new phone service gave her the ability to get data anytime and anywhere. It didn't show her driving, but the idea that rankled Reggie was: "I was that guy. I was the guy on the phone all the time." He thought: Why not put the phone down and get in the moment "and enjoy yourself a little."

Years ago, public safety advocates concerned with distracted driving focused on the marketing of cell phone companies, which were pushing the idea of the "car phone." Now, safety advocates say, the most potentially destructive marketing and product design comes from a different industry, the carmakers. In recent years, they've made a big push toward creating "infotainment" systems that feature touchscreens in the dashboard to allow navigation, access to all kinds of media, including satellite radio and playlists, and even some video content.

The latest are systems that allow voice-activated commands so a person can stay connected to all kinds of Internet functions while driving. For instance, a driver can dictate a text or an email or a Facebook update. High-end BMWs already allow drivers to dictate emails or send texts, and the Chevrolet Sonic lets drivers compose texts to an iPhone. More than half of all new cars will integrate some type of voice recognition by 2019, according to the electronics consulting firm IMS Research.

In early 2013, General Motors announced partnership with AT&T to bring Wi-Fi hotspots to cars. The carmaker's chief executive, Daniel Akerson, formerly a telecommunications executive, was quoted as telling Reuters that the business potential was big.

"I have grandchildren that have only grown up in a world with smartphones," Akerson said. "With a 4G pipe into a car, you can change the business model almost entirely—for example, what happens if when the logo shows on your screen, it says 'brought to you by Allstate'? How many times is that going to pop? And how much can you get from Allstate?"

In April 2013, for example, Audi put out a press release announcing a partnership with T-Mobile, the cell phone provider, and heralding: "the industry's most competitively priced in-vehicle data plan—providing Wi-Fi for up to 8 connected devices, a full suite of Google search and mapping services, and Sirius XM traffic information through Audi Connect." The press release touted a new video called Beach Day, which, as the company explained: our driver quickly and easily plans an outing on the go while chatting with her mother in-vehicle. "The video shows how Audi Connect can soothe anxieties associated with day-to-day travel by enabling users to make plans on-the-fly, calculate the distance to gas stations, access real-time weather info and locate nearby points-of-interest."

The carmakers have said the systems are safer because they keep the driver's hands on the wheel and their eyes on the road.

There's a catch. A groundbreaking study by Dr. Strayer in 2013, which relied on assistance from Dr. Gazzaley, among other top neuroscientists, found that the voice-activated commands are actually more dangerous for drivers than talking on a phone, even a handheld one. The reason is that, even though a driver's hands may be on the wheel and their eyes on the road, their minds are elsewhere—captured by the technology. The act of talking to these systems, even when they work seamlessly (meaning: no dictation problems), engages speech processing, as well as the parts of the brain involved in planning the construction of the command. As a result, drivers are significantly less likely to see an obstacle, stay focused on the road, and more likely to suffer inattention blindness.

The carmakers have good reason for wanting to sell these gadgets: consumers say they want them. That's a big thing for auto companies faced with the problem that people are tending to hold on longer to their cars before buying new ones. Offering new technology is a way to get people into the showroom to consider a new purchase, says Ronald Montoya, the consumer advice editor for Edmunds.com, a car research firm. The technology "is getting people to try new cars rather than stick with buying used cars," he says.

According to Montoya, the next trend is "integration with apps that are on your phone—having them display onto the screen itself."

Speaking of screens, as they were getting commonplace in cars, they were growing in size, too. In the Tesla, the hip electric car, there was a seventeen-inch screen, which Montoya described as "essentially a 17-inch-high iPad that controls everything in the car." It's "cool," he adds, but as a consumer advocate he'd prefer something more direct, like an old-fashioned push button that lets a driver control air-conditioning, heat, or the radio so "you don't have to dig through a couple of screens in the menu. Even turning up the volume is a touchscreen, which is not as responsible as you just turning a dial."

David Teater, a former auto industry executive, has concerns about these innovations. He knows the risk of driver distraction firsthand. His own son was killed by a young woman who was driving while talking on her phone, an event that turned Teater into one of the foremost advocates for curbing distracted driving, particularly cell phone use by drivers.

He says that all the glorification by automakers of technology use by drivers "normalizes" the behavior.

"The more the automakers make these standard in a vehicle, the more this is what you do."

He likens the marketing of devices in the car to the cigarette industry. "It's almost like back in the day of the tobacco industry. The automakers are saying: 'We need more research, we need more research,' and they keep building all the technology into vehicles, which is normalizing the behavior and making it harder and harder down the road to say: 'We're killing all these people, we need to stop.'"

SOME IN THE TECHNOLOGY industry were acknowledging the issues of distraction. In an article published in July 2012, in the *New York Times*, I outlined how the idea that devices, sites like Facebook, and video games could be addictive was being embraced by, of all people and places, the leaders in Silicon Valley. The conversation was happening inside Facebook, and Google, and at companies making the huge telecommunications infrastructure delivering that technology, like Cisco. They were talking, for the first time in any focused way, about the role of dopamine, how the brain might be impacted by the technology—the constancy of pings and demands for attention, how it might even be addictive.

People at those companies began experimenting with different approaches—such as meditation or scheduling technology breaks—to keep employees focused and productive, not scattered and creatively drained.

"If you put a frog in cold water and slowly turn up the heat, it'll boil to death—it's a nice analogy," Stuart Crabb, director of learning and development at Facebook, told me for the story. People "need to notice the effect that time online has on your performance and relationships."

He and hundreds of other people from major companies like Twitter, eBay, and PayPal were getting involved in the Wisdom 2.0 conference, aimed at helping people find downtime, get offline. In the story, I also spoke to Richard Fernandez, an executive coach at Google, who said, "Consumers need to have an internal compass where they're able to balance the capabilities that technology offers them for work, for search, with the qualities of the lives they live off-line."

"It's about creating space, because otherwise we can be swept away by our technologies."

But the article pointed out that these companies were still predicating their business models—making money—on keeping people connected all the time.

ON THE POLICY FRONT, the Department of Transportation in 2010 did pilot tests in two cities, Hartford, Connecticut, and Syracuse, New York, in which they did public-service announcements about the public-safety dangers and risk of criminal penalties from illegal phone use by drivers. The campaign was coupled with heavy law enforcement activities: police in Syracuse gave out 9,587 citations, and in Hartford, 9,658.

The federal agency tracked behavior before and after the campaigns and claimed to find a decided difference in cell phone use by drivers. In Syracuse, the agency said, phone use and texting dropped by a third, while in Harford, the agency found a 57 percent drop in handheld phone use, while texting dropped by three-quarters. The agency pledged to do more such campaigns, which it called "Phone in One Hand, Ticket in the Other" and was modeled after the seat belt effort "Click It or Ticket."

Harsha says that only so many lessons can be drawn because of the intensive resources the campaign demanded in Syracuse and Hartford, including the use of "spotters" who relayed what they saw to other officers down the road. Not an inexpensive or simple solution, she says. "I'm not convinced that enforcement and education alone will solve the problem, especially when enforcement is so difficult."

By the middle of 2014, laws banning texting had grown in popularity but were neither uniform nor universal, according to the National Conference of State Legislatures. For instance, the group reports, Arizona, Montana, South Carolina, and Puerto Rico have no bans, while Missouri, Oklahoma, Texas, and Mississippi only ban novice drivers from texting. And a handful of other states have bans that are "secondary" laws, meaning a driver can't be pulled over merely for texting and driving, but only cited if the driver is doing something else wrong (a primary offense). Only twelve states and the District of Columbia have bans on the use of handheld phones by drivers. But around the country, even where there are laws, texting continues basically unabated, and so does the disconnect: Drivers say overwhelmingly that it is risky, but they can't stop themselves from doing it, or don't want to. In the summer of 2013, Kars4Kids, a car-donation charity, did a survey of drivers that illustrates the wide gap between behavior and attitude. The survey found that 98 percent consider texting and driving to be dangerous, but 43 percent of those surveyed read texts, and 30 percent sent them. There was another interesting statistic: 46 percent of passengers said they had texted while the driver was driving. In other words, rather than focus on the driver, or help the driver focus on the road, the passengers were more interested in their devices. "And the carnage keeps piling up," Dr. Strayer told me in late October of 2013, seven years after Reggie's accident. He was getting ready to fly from Salt Lake City to San Francisco to testify in a deposition of a lawsuit in which a woman was using her phone and ran over and killed a pedestrian. He said he gets a call once a week to get involved in a legal case.

When Reggie got behind the wheel on September 22, 2006, he

could perhaps have made the argument he'd not heard much about nor understood the risks of texting. He'd be the first to say that didn't make his behavior excusable; he'd been taught to be an attentive person, and driver. It was raining as he crossed through the mountains, the sun just rising, on the commute that would forever alter his and others' lives. The people responding to the latest surveys can't claim they are ignorant of the risks of texting. They own up to it. They know the truth, if they can only get themselves to listen to it.

To Ms. Harsha, progress made by the safety community runs smack into a case of collective denial, reinforced by powerful marketing messages urging people to stay connected. "The culture is: 'It's not me, it's you. I'm the good driver.'

"Part of education is getting people to face up to their own behavior."

IN MARCH OF 2014, Terryl got a big honor. She was appointed to the Utah State Board of Education by Governor Gary Herbert. In choosing Terryl, the governor lauded her efforts as a victim's advocate and said: "Her experience will greatly benefit the families of Utah."

THE NEUROSCIENTISTS, IN ADDITION to pursuing individual efforts to understand the impact of technology use on the brain, were stepping up their collaborative efforts. In 2013, Dr. Gazzaley, Dr. Strayer, Dr. Atchley, and other leaders in the field earned a joint grant to study the power of nature to rebuild tired brains, restore them. They met together in the spring of 2014 to embark on this new quest.

"People say: 'I put down the phone and hang out on the beach or go camping or rafting or fishing and I feel so rested.' We're trying to prove the biological basis of that," Dr. Strayer says, adding that the research is more in the spirit of Henry David Thoreau than of Donders and Helmholtz, Broadbent and Treisman and Posner. "Too much technology," Strayer argues, "can corrupt the soul."

Dr. Atchley takes a more scientific approach to the same question, connecting the dots between spending some time away from technology and finding some truth, making clearer-headed decisions.

"To make a choice, you need frontal lobes active, and need few enough competitors in other parts of the brain so that you can engage systems to make a decision," he says. "You have to have the capacity.

"An ethicist would say we all have a choice, but a brain scientist will tell you that choice starts in the brain."

AUTHOR'S NOTE

I MET REGGIE IN August 2009, shortly after he'd gotten out of jail and not long after I had found myself in the middle of the rarest of journalistic maelstroms. It had begun the previous December—right about the time that Reggie was deciding to cop a plea—when I had a conversation with Adam Bryant, an editor and close friend at the *New York Times*.

We were talking about the concept of driving while using a cell phone. It was, on one level, just another conversation. Adam and I chewed on everything that came across our minds—chats that lived in the gray area journalists themselves inhabit that involves remarking on everything in the world in a casual way while also wondering aloud if the things we'd observed merit a story.

Among the topics Adam and I had often discussed was the impact of technology on behavior and society. We were looking for "disconnects," places where things didn't work as advertised, or where people made assumptions or presumptions—say, the inherent good of technology—that weren't necessarily borne out. Silicon Valley, which I was covering, was a fruitful area for lots of journalists trying to get at the underbelly of this explosion in personal communications.

Part of my curiosity was also personal. I was watching how technology was altering my own behavior. I could see that I was spending a lot of time with media, changing my communications patterns, feeling a tick of anxiety when away from my device, using it to escape from uncomfortable moments or boredom, even when behind the wheel. It wasn't so much that I saw this technology as bad or good, but simply as powerful.

In 2003, I'd written a piece for the Sunday business section called "The Lure of Data: Is It Addictive?" I started the story with a venture capitalist who was sitting at a conference he'd paid $2,000 to attend, but he said he wasn't paying attention at all to the goings on because he was simultaneously on his laptop and his phone. "It's hard to concentrate on one thing," he said in the story. "I think I have a condition."

I was also observing, professionally and personally, the extraordinary power of technology, as a productivity tool and a tool for liberation and creativity. Thanks to technology, I didn't have to work in New York, but could be in San Francisco, which allowed me to write about Silicon Valley, be immersed in it. I could talk to my editors daily, hourly, whenever, and file my stories from my home office or on the go. I once wrote a front-page story about the legal arguments made in a lawsuit against Napster, the music-file-sharing company, while sitting on the floor outside the courtroom on the Ninth Circuit in downtown San Francisco, and then called the story in to my editors by cell phone.

I discovered that technology, by providing such flexibility, could make me not just more efficient but, in a way, more creative. I could work when and if the muse hit. I could also take breaks more easily, say, jet off to the gym during lunch, and take my phone to make sure I wasn't missing something. At the time, I didn't realize how much more effective that could make me, but something visceral spoke to me about the power to have more control over my life.

This piece, this liberation, was its own mixed blessing. I wasn't putting so fine a point on it at the time, but I was beginning to wrestle in my own life with questions of efficiency and opportunity. So what if we could do more work? So what if we could stay in touch all the time?

Should we? It was thinking that would evolve in no small part thanks to my interactions with Reggie.

At the time, back before I met him, these conflicting forces helped fuel and indulge a muse that bowled me over in 2004. I wrote a novel called *Hooked: A Thriller About Love and Other Addictions*, that posited about the addictive powers of technology and whether it might change our realities and perceptions. It was more science fiction than science, but I did borrow from some of the fledgling science I'd discovered in the story I wrote for the paper a year earlier. Technology was the culprit in the book, and no small companion and collaborator in helping me write and then publicize the book.

WHEN ADAM AND I started talking in December 2008, we decided I'd start looking into what Adam calls a "smart dumb question" regarding one particular issue: Why are people using their phones while driving? Isn't that dangerous? Shouldn't they know better?

The following spring, I started reporting the story as I took on other daily assignments. It was something of a right of passage for me as a reporter to be asked to look into something in some depth, to be allowed the time to investigate something that might not bear fruit. I took my time, and I learned a lot.

I made a trip to Utah and spent time meeting with Dr. Strayer, who'd clearly distinguished himself as a leader in the field. While there, I also talked to the family of Lauren Mulkey, the young woman who was seventeen years old when she'd gone out with friends for Saint Patrick's Day and got killed by a distracted driver. When I returned from that reporting trip, I sent a memo to Adam that explained much of Dr. Strayer's research, and that summarized other things I'd discovered, like the growing research on texting (not just in Dr. Strayer's lab but at the Virginia Tech Transportation Institute), and I wrote to him about what seemed like a somewhat startling discovery: allegations that the federal government had known about the risks of using a phone while driving well back around 2000 and had essentially buried that information for political reasons.

My memo proposed we write a story "that weaves one very human tale of a fatal accident along with the emerging science and, pointedly: the evidence that the government has been slow to release clear evidence of dangers of cell phones in cars. In fact, I can almost see a parallel structure in this story: our protagonists not paying attention to the road while the government is not paying attention to the evidence."

As my reporting continued, I found other stories—tragic, terrible tales. One of them was the story of Linda Doyle, who, in September 2008, had been driving in Oklahoma City, not far from her house, when she was killed by another motorist, Christopher Hill. Hill had been talking on his phone and had run a red light at high speeds.

On July 19, 2009, the paper ran on its front page a long, ambitious narrative that weaved the collision of Hill and Doyle into a story about the science, politics, and policy around distracted driving. It was so ambitious because we thought this was it, one story aimed at putting a stamp on this issue. Underscoring how much we invested in that story was an added element of presentation: The *New York Times* produced a video game, a sort of driving simulator, that people could play to test their own abilities to multitask.

We thought the article might have legs, but we were blown away by what happened. The story just took off. It got picked up by lots of media, and was viewed and emailed in big numbers on the *New York Times* website. It received 655 comments, which, within even a year, would not seem such a big deal, but at the time it was astronomical. People were talking about how the story got at something that they were feeling and experiencing acutely in their own lives.

Because I'd invested so much in reporting in the prior six months, I had a lot of other story ideas in my notebook. We went with the momentum, getting key encouragement and wisdom from Glenn Kramon, the paper's editor for large enterprise projects. On July 21, we published a mini-investigative piece that showed that the Department of Transportation, a federal agency, had in 2003 known about and withheld compelling research showing the deadly risks of multitasking behind the wheel. The story explained that the federal agency didn't publish its data—including estimates that cell phone use by drivers caused around 955 fatalities and 240,000 accidents over all in 2002—because of fears of angering Congress. (Congress had previously admonished the Department of Transportation to avoid certain political issues.)

In the story, I quoted Clarence Ditlow, director of the Center for Auto Safety, as saying: "We're looking at a problem that could be as bad as drunk driving, and the government has covered it up."

On the next Sunday, July 27, we published a story about research from the Virginia Tech Transportation Institute that showed texting by truck drivers increased risk of crash or near crash by twenty-three times. The story also talked about Dr. Strayer's research. At the time, fourteen states had banned texting while driving, including Utah.

The reaction to the articles was something my editors said they'd rarely seen, and whereas we thought I'd do one or two stories on the subject, now my editors wanted to know what else was in my notebook.

"There's a guy I've heard about in Utah," I told Adam. "His name is Reggie Shaw."

TERRYL PICKED ME UP at the airport. She was as perky as she'd been on the phone, and at once friendly and intense. What most stuck out in our conversation as we drove up I-15 was something totally unrelated to the story. She talked about how she wanted to encourage young Mormon women to go to college before they got married, or at least to find some sense of purpose to go along with having and raising a family. This seems pretty basic, but she was from Logan, where people got married young and had kids, and that was the point. She wanted to shake up her environment and challenge assumptions.

I also was struck by the crash site. It was the scene of horrific violence, and yet it was surrounded by abject beauty. You could see how Brigham Young would've come over the mountains into Utah and thought: *This is it.*

In the Cache County prosecutor's office, I met Leila and Jackie. Leila was still nearly inconsolable with grief.

I first met Reggie in Tremonton, I believe. He wouldn't go to the crash site. It was too painful. I did drive back with him to Salt Lake City, and he told me the story about the young woman he'd been on a date with who had the newspaper article about him taped to her mirror. Reggie was self-effacing, kind, open, and carrying a deep, open wound.

The article about Reggie ran on August 29, 2009.

IN THE SPRING OF 2010, the paper and I were awarded the Pulitzer Prize for National Reporting, a significant validation that we'd illuminated a widespread hazard, and crediting us with "stimulating widespread efforts to curb distracted driving"; as part of our supporting evidence in the paper's nomination, we noted that *Webster's* dictionary had named "distracted driving" its word of the year in 2009. There was another Pulitzer Prize given that year, to the *Washington Post*, that had me thinking about the bigger issues at stake with distraction. The spectacular feature story focused on the trial of a man who had been so distracted by a cell phone call that he'd left his son strapped in a car seat in a sweltering car. The car was not moving. The man was not driving. The boy spent nine hours strapped inside the car, and died.

That spring, I embarked on a series of stories called "Your Brain on Computers" about how heavy technology use impacts the brain. I learned more about the addiction side, began to understand the connections between some of these ideas, and kept trying to understand the central paradox: We all know that texting and driving is dangerous, but we do it anyway. *Why?*

Among the things I learned was how widespread the potential had become for distraction. In one story, I explored how doctors are getting distracted, a story that included one of the most harrowing anecdotes I'd heard about in all my reporting on a subject: A neurosurgeon in Denver was accused of making personal calls during surgery and, owing to distraction, left the patient partially paralyzed. It was a particularly horrific outcome but not a totally singular example of electronic distraction among doctors, nurses, and key medical technicians.

A medical journal called *Perfusion*, focused on cardiopulmonary bypass surgery, did a survey and found that 55 percent of technicians monitoring the bypass machines said they'd talked on their cell phones during surgery and half had texted. They did that even though about 40 percent said talking on the phone was "always an unsafe practice," and around half said the same of texting. As I noted in my story, the study's authors wrote: "Such distractions have the potential to be disastrous."

ALL THAT REPORTING GAVE me a backdrop and experience to approach this book. But it was only that: backdrop. This book was, except for a handful of instances where I quoted previous articles, reported entirely from scratch.

The reporting for this book consisted of: extensive phone and in-person interviews with dozens of people; police and law enforcement reports; historical documents, some first-source and some accounts collected by reputable sources, like museums; use of audio and video recordings and of transcripts from law enforcement interviews and of court proceedings. In particular, these transcripts and recordings were used to reconstruct witness interviews, court hearings, and legislative proceedings. Quotations were taken directly from the recordings or transcripts. Without exception, I tried to stay completely true to the spirit of the event. So, too, with the letter; only in rare cases did I change a word or use ellipses to try to avoid confusion. I summarized sections that were redundant or ancillary.

Other crucial documentation included handwritten notes and personal records made at the time of the events from participants, including Gaylyn White, Reggie's therapist, and Jon Bunderson, Reggie's lawyer. I was allowed access to these records only because Reggie gave his permission to allow them to be shared with me, or turned over.

Another key allowance was made me by the court and Judge Willmore. He unsealed the record of this case, which had been previously sealed. Reggie also acceded to this. It enabled me to see videotapes of key hearings, as well as read certain crucial documents. I also received both official and unofficial documentation (Terryl's handwritten notes) from the prosecutor's office, and provided by Terryl.

As noted, there are a handful of places in the book where I cite reporting from other media, notably the *New York Times*. In many cases where I made such a citation, I was relying on a story I had myself written for the paper, reflecting my own familiarity with the material. In such cases, I did not necessarily write in the book that I had written the *Times* article. That is simply because I did not wish to interrupt the flow of the story with an introduction of my role in the reporting.

This powerful trove of documentation, in totality, not only gave me a concrete base of reporting. It also served to reinforce and provide context and guidance for what was the most significant source of reporting: the extraordinary cooperation of the people in this book. One person after the next shared their stories, perspectives, motivations, hopes, and fears. These countless conversations, some by phone, some in-person, involved much soul-searching, sometimes tears, and laughter. In that way, this book represents a collective expression, even exhalation, of emotion that I feel very fortunate to have been able to record. Broadly, I offer my deepest thanks to everyone who saw fit to share of themselves.

Memories can obviously be unreliable, and accounts biased. To create the fairest possible account, I sought to use more than one source to document events. By way of example, when describing Reggie's interaction with his counselor, Gaylyn, I matched his account with hers, and with the documentation from the sessions. That same principle applied to meetings among the prosecutors, or when Reggie and his family met with Bunderson. Much more often than not, the accounts of various parties were highly consistent; people might have had different perspectives of an event—say, how they felt at the crash site—but the accounts of what was said, and by whom, rarely differed or by much. There is also no substitute for getting to know people over time, which is a luxury I was afforded given that I started meeting the key people in this story in 2009 and remained in contact until the final reporting in the spring of 2014.

Some of the most powerful and personal experiences came only from a single source and, perhaps, necessarily so. For instance, Don Linton's revelations about the sexual abuse he suffered came from him. In numerous lengthy, emotional conversations, I came to appreciate the veracity of his experiences and emotions and could find no reason for him to exaggerate. If anything, it was a risk for him to share his past. I admire his courage. That is true, too, for Tony Baird, whose tearful discussions of his childhood accident were courageous and raw, and I am indebted. Same goes for David Greenfield and his own challenges growing up. In the case of Terryl, and her childhood, I felt I needed to apply a particularly high standard because she is such a central part of the story. Much of her account of her childhood comes from her. In the spirit of the highest standards of journalism, I matched her accounts with her diary of her childhood, which she provided. I also corroborated her stories by interviewing people who were close at the time with Terryl and her family. I did not do these things because I doubted Terryl in any way. In fact, as I came to know her, I discovered someone with a sense of ethics and dedication to truth rivaling those I find among the finest journalists I have come to know. Terryl's mother declined to be interviewed, though I obviously tried to reach out to her—through direct email, and in overtures made through Terryl and her little brother Mitchell. Mitchell, who graciously agreed to be interviewed, asked that his current situation be excluded from this story. Through Terryl, I was privy to emails written by her older brother Michael that corroborated some of the events Terryl described. In some cases, it was difficult to put together the precise dates and chronology of events from her childhood and I made my best judgment by combining the interviews with Terryl and her friends and family's friends, with the diary accounts, and with some photographic evidence.

With one exception, which I will note a bit farther down, there can be no fiercer advocate than Terryl Warner.

I received extensive cooperation from Jackie Furfaro and Leila O'Dell. They invited me into their homes, and, even more powerfully, into their thinking and emotions. I am deeply indebted to these two proud, strong, courageous women, and to their families. They have experienced loss no person should ever experience, and they have persevered.

From the first time that I met Reggie, I was astounded at his openness. He told how he felt and what he'd done, and he told me when he couldn't figure out what he'd done, or why, or how. Reggie was and is an open book. And even though he can be shy, he does not crave attention and would rather be doing a lot of things besides talking. His openness reflects what I can only describe as a shame and sorrow and desire to redeem himself that goes deeper than I've personally witnessed in any person under any circumstance. This is why he is the fiercest advocate I can imagine. He desperately wants no person to do what he has done. He has gone down on his knees and begged. But I hope he will rise up, and heal. Virtually every person I met on this journey—from his own family, of course, to Terryl, to Judge Willmore—want to see Reggie find peace. I share the same hope for Reggie.

Reggie, I hope that, in some small way, the contribution you have made to this account adds sufficiently to the portfolio of your testimony so that you can forgive yourself.

ACKNOWLEDGMENTS

FOREMOST, THIS BOOK WOULD not exist without the cooperation of dozens of people who shared their lives, often in intimate and exquisite detail. To all of these people, I offer you my most humble thanks. You put your faith in me, and this process, and I am indebted beyond my ability to express.

Specifically, thank you to Leila and Megan O'Dell and to Jackie Furfaro and your family for giving so generously of your time, and your innermost feelings, to allow the chronicle of this tragedy.

My thanks to John Kaiserman for a patient and detailed recounting.

Thank you to the men and woman of Utah's law enforcement community, including Bart Rindlisbacher, Scott Singleton, and Kaylene Yonk, who graciously shared their process, expertise, and much about their own lives. Prosecutors Scott Wyatt, Tony Baird, and Don Linton patiently outlined the legal processes, their approach, and, in the case of Mr. Baird and Mr. Linton, deeply personal histories that, crucially, explained some of the ways their own experiences helped shape the outcome of the case. Thank you.

Thank you to Judge Thomas Willmore for your time, personal reflections, and terrific insights about the law. And thank you for similar insights and personal accounts from Utah legislators, notably Stephen Clark and Carl Wimmer. Thank you to former ambassador and governor Jon Huntsman, Jr., and to former congressman and secretary of transportation Ray LaHood, for your insights and leadership on distracted driving. And thanks to safety advocates David Teater, Barbara Harsha, and Bill Windsor.

Thank you to Jon Bunderson, Reggie's attorney, and Gaylyn White, Reggie's counselor, who provided time, insights, and, with Reggie's permission, access to your notes.

Thank you to the numerous top scientists who painstakingly explained difficult concepts and in some cases their personal stories: Dr. Daniel Anderson, Dr. Ruthann Atchley, Dr. Daphne Bavelier, Dr. Nicholas Christakis, Dr. Susan Forward, Dr. Marc Galanter, Dr. Daniel E. Lieberman, Dr. Alan Mackworth, Dr. Earl Miller, Dr. Michael Posner, Dr. Marc Potenza, Vicky Rideout, Dr. Gary Small, and Dr. Jason Watson. My deep thanks to Dr. Anne Taylor Treisman, a pioneer, who illuminated the early days of neuroscience through a prism of her own experiences. A special thanks to four scientists who made complex, cutting-edge science accessible through their lives and work: Dr. Paul Atchley; Dr. Adam Gazzaley (and to Jo Fung), Dr. David Greenfield, and Dr. David Strayer. Simply, this book would not have happened without your cooperation. A key contributor to my understanding of the science, and a world-class scientist, Dr. Clifford Nass, died far too young. We mourn your loss.

My thanks to Terryl Warner and to her family, husband Alan, and to Jayme, Taylor, Allyssa, and Katie. You all shared well beyond the call.

My thanks to Mitchell Danielson for making the time, and entrusting me with your story.

Thank you to the people of Tremonton, including Dallas Miller, Jason Zundel, and to Van and Lisa Park.

I wish to offer my deepest thanks to Mary Jane and Ed Shaw. You opened your lives and home, shared our hopes and fears, and the rawest emotions from a terrible tragedy. Thank you to Phill Shaw.

Thank you to Reggie. Per my author's note, you have laid yourself bare to me, and the world. As an author, as a citizen, I say: thank you; you have suffered enough, and done enough.

Thank you to the extraordinary team at William Morrow/Harper Collins, led by publisher Liate Stehlik, a true author's friend. Thank you to my superb editor, Peter Hubbard, and to the creative and energetic marketing and publicity team, Shelby Meizlik, Andy Dodds, Tavia Kowalchuk, and to Julia Black and Adam Johnson. Thank you to the national sales team for terrific support, and to Trina Hunn for great care.

Thank you to my great friend and agent, Laurie Liss, at Sterling Lord Literistic.

Thanks for terrific research and counsel to Sophie Egan, Lois Collins, and Sean Hales.

Mom and dad, you are tirelessly supportive. Thank you.

I can never repay the friendship and brotherhood of Bob Tedeschi, a sage who counseled me with compassionate wisdom in this process as he has in many adventures.

My undying thanks and all my love to my beautiful wife, Meredith Barad, a brilliant sounding board and patient listener, and to our magnificent children, Milo and Mirabel. I love you, always.

INDEX

cell phones (*continued*):
  i-mode networking standard for, 26
  lobbyists for, 190
  lying about use of, 283
  and multitasking, 86, 125–26, 229
  news apps on, 218–19
  risk analysis of, 25–26, 276–77
  smartphones, 26, 198
  and social connection, 85, 215, 282, 369–70
  teen drivers' use of, 86
  and texting, 26, 165, 271
  and texting, laws against, 25, 189–91, 212, 282–83
  visual, manual, and cognitive demands of, 126, 217, 229
central nervous system, 103
Cherry, Edward Colin, 103
chocolate cake, and decision studies, 219, 353
Christakis, Nicholas A., 215
Christensen, Nate, 330
Cisco, 373
Clark, Stephen, 249–50, 295–97, 299, 300–301, 304–5, 332
Clayton, Elder, 235
cocaine, 195
cocktail party effect, 62–63, 69, 103
cognitive neuroscience, 100
cognitive psychology, 103, 104
Common Sense Media, 218
communications:
  attention controlled by, 108
  and computers, 5, 67–68, 85–86, 103, 141, 143–44, 323
  email, 70, 199
  instant messaging, 85
  networks of, 123
  online journals, 70
  smartphones, 26, 198
  and social connection, *see* social connection

telecommunications, 103, 141, 143, 373
  texting, 70
comorbidities, 197–98
Computer History Museum, Silicon Valley, 26, 67
computers:
  and children, 86
  and communications technology, 5, 67–68, 85–86, 103, 141, 143–44, 203, 323
  development of, 4, 65
  evolution of, 67, 141–42
  Moore's law and Metcalfe's law, 5, 143–44, 352
  punch cards used in, 67
concentration, brain studies of, 33
Counting Crows, 62, 68
Crabb, Stuart, 374

Daines, George, 207, 209, 213, 223, 225–26, 227, 230, 233, 265, 267
Danielson, Byron Lloyd "Danny":
  abusing his family, 49–54, 72–73, 74–75, 78–80, 343, 351
  drinking, 50, 52, 72–73, 74
  and guns, 49, 50, 54, 73
  and Terryl's adulthood, 134
Danielson, Kathie, 49–54, 73, 74–76, 79, 80, 152, 342–43
Danielson, Kerryl, death of, 50
Danielson, Michael, 49–54, 73, 75, 80, 134, 152, 343
Danielson, Mitchell, 51–54, 73, 75, 78–80, 343
Danielson, Terryl:
  biological father of, 76, 80, 152–53
  and Danny, 49–54, 72–73, 74, 75, 78–80, 347, 351
  escapes for, 74, 75
  jobs held by, 74, 77
  marriage of, *see* Warner, Terryl
  in school, 76–77, 78
DARPA (Defense Advanced Research Projects Agency), 141

About the Author

MATT RICHTEL reports for the *New York Times*, covering a range of issues, including the impact of technology on our lives. In 2010 he won the Pulitzer Prize for National Reporting for a series of articles that exposed the pervasive risks of distracted driving and its root causes, prompting widespread reform. He is the author of four novels, including, most recently, *The Doomsday Equation*. A graduate of the University of California at Berkeley and the Columbia Journalism School, he is based in San Francisco, where he lives with his wife, Meredith Barad, a neurologist, and their two children.